AF784913

Modern Control Techniques for the Processing Industries

CHEMICAL INDUSTRIES

A Series of Reference Books and Textbooks

Consulting Editor
HEINZ HEINEMANN
Heinz Heinemann, Inc.,
Berkeley, California

Volume 1: Fluid Catalytic Cracking with Zeolite Catalysts, *Paul B. Venuto and E. Thomas Habib, Jr.*

Volume 2: Ethylene: Keystone to the Petrochemical Industry, *Ludwig Kniel, Olaf Winter, and Karl Stork*

Volume 3: The Chemistry and Technology of Petroleum, *James G. Speight*

Volume 4: The Desulfurization of Heavy Oils and Residua, *James G. Speight*

Volume 5: Catalysis of Organic Reactions, *edited by William R. Moser*

Volume 6: Acetylene-Based Chemicals from Coal and Other Natural Resources, *Robert J. Tedeschi*

Volume 7: Chemically Resistant Masonry, *Walter Lee Sheppard, Jr.*

Volume 8: Compressors and Expanders: Selection and Application for the Process Industry, *Heinz P. Bloch, Joseph A. Cameron, Frank M. Danowski, Jr., Ralph James, Jr., Judson S. Swearingen, and Marilyn E. Weightman*

Volume 9: Metering Pumps: Selection and Application, *James P. Poynton*

Volume 10: Hydrocarbons from Methanol,
Clarence D. Chang

Volume 11: Foam Flotation: Theory and Applications,
Ann N. Clarke and David J. Wilson

Volume 12: The Chemistry and Technology of Coal,
James G. Speight

Volume 13: Pneumatic and Hydraulic Conveying of Solids,
O. A. Williams

Volume 14: Catalyst Manufacture: Laboratory and Commercial Preparations, *Alvin B. Stiles*

Volume 15: Characterization of Heterogeneous Catalysts,
edited by Francis Delannay

Volume 16: BASIC Programs for Chemical Engineering Design,
James H. Weber

Volume 17: Catalyst Poisoning,
L. Louis Hegedus and Robert W. McCabe

Volume 18: Catalysis of Organic Reactions,
edited by John R. Kosak

Volume 19: Adsorption Technology: A Step-by-Step Approach to Process Evaluation and Application, *edited by Frank L. Slejko*

Volume 20: Deactivation and Poisoning of Catalysts,
edited by Jacques Oudar and Henry Wise

Volume 21: Catalysis and Surface Science: Developments in Chemicals from Methanol, Hydrotreating of Hydrocarbons, Catalyst Preparation, Monomers and Polymers, Photocatalysis and Photovoltaics
edited by Heinz Heinemann and Gabor A. Somorjai

Volume 22: Catalysis of Organic Reactions,
edited by Robert L. Augustine

Volume 23: Modern Control Techniques for the Processing Industries, *T. H. Tsai, J. W. Lane, and C. S. Lin*

Additional Volumes in Preparation

Temperature-Programmed Reduction for Solid Materials Characteristics, *Alan Jones and Brian McNicol*

Catalytic Cracking: Catalysts, Chemistry, and Kinetics, *Bohdan W. Wojciechowski and Avelino Corma Canós*

Chemical Reaction and Reactor Engineering, *edited by J. J. Carberry and A. Varma*

Filtration: Principles and Practices, second edition, *edited by Michael J. Matteson and Clyde Orr*

Modern Control Techniques for the Processing Industries

T. H. TSAI
J. W. LANE
C. S. LIN
Tenneco Inc.
Houston, Texas

Marcel Dekker, Inc. New York and Basel

Library of Congress Cataloging in Publication Data

Tsai, T. H. (Thomas H.), [date]
Modern control techniques for the processing industries.

(Chemical industries ; v. 23)
Includes bibliographies and index.
1. Chemical process control. I. Lane, J. W. (James W.), [date] II. Lin, C. S. (Cheng S.), [date] III. Title. IV. Series.
TP155.75.T83 1986 660.2'81 86-2112
ISBN 0-8247-7549-X

Copyright © 1986 by Marcel Dekker, Inc. All Rights Reserved

Neither this book nor any part may be reproduced or transmitted in any form or by any means, electronic or mechanical, including photocopying, microfilming, and recording, or by any information storage and retrieval system, without permission in writing from the publisher.

Marcel Dekker, Inc.
270 Madison Avenue, New York, New York 10016

Current printing (last digit):
10 9 8 7 6 5 4 3 2 1

Printed in the United States of America

To Cathy, Lynn, and Cheryl

Preface

The advancement in digital control technology has drastically changed the instrumentation and control industry. As of 1984, there were over 50 digital control system suppliers who offered control systems with a wide variety of capabilities, as compared to fewer than 10 control system suppliers whose products were not very different in terms of functional capabilities just 10 to 15 years ago. Today's control systems offer a wide range of built-in functions, and these functions should result in superb performance when they are properly utilized. Unfortunately, most of the digital control systems that exist today were installed as analog controller replacements, so the advantages of these control systems have only been recognized as possessing qualities of higher reliability and more extensive operator communications when compared to conventional control systems. Many people have not implemented control functions more advanced than those implemented by conventional controllers in the past.

We believe the major reason for the "underutilization" of digital control systems is the lack of engineers adequately familiar with this field. According to the results of a survey of 3,599 chemical engineers conducted by *Chemical Engineering* (September 19, 1983), the engineering course generating the greatest dissatisfaction during the past decade has been the course on process control. The major complaint among young graduates was that too much emphasis was being put on Laplace transforms, too little on controlling processes.

With little or no formal education on controlling processes, most young chemical engineers have learned their practical process control knowledge from "old-timers." Unfortunately, many old-timers learned from the application of single-loop analog controllers, and may not have much knowledge in the application of computers. Their common approach for designing a control system is to use a digital system to simulate a proven system using analog controllers. The young engineers, although knowledgeable about digital computers, are being led to follow similar approaches, because these are the only systems proven in the field.

We have observed that the equipment in many plant control rooms these days looks more like that on spaceships in science fiction movies than conventional panel boards and consoles, but behind all those color CRT displays and the flashing lights, there is nothing more sophisticated than proportional-integral-derivative (PID) control algorithms. We believe that this is a great waste of resources.

On the other side of the spectrum, we have read many publications about advanced control techniques, particularly about developments in new control algorithms from many academic institutions. Unfortunately, few of these developments have been utilized in actual plant control. This also seems to be a waste of resources.

We decided to write this book in an attempt to provide a reference on the application of computer-based control systems, emphasizing the not-so-conventional, but practical approaches. We hope this book will give old-timers in the instrumentation and control industry some new concepts to consider, and will give young engineers a general guideline on how to put their theoretical training into practical use.

This book is divided into two parts, "Process Control Fundamentals" and "Process Control Applications." The first part provides a review of existing techniques necessary to analyze a process and to design a control system, while the second part discusses application concepts based on our own experiences.

To understand this book fully, readers should have college-level mathematics and a working knowledge of plant operation, instrumentation, and control systems.

We are grateful for the encouragement received from Dr. Maurits Dekker to start this project. Special thanks also go to Mrs. Linda Korbut for her tireless typing and proofreading of this manuscript. We would also like to acknowledge the generous assistance we received from our good friends Mr. John E. Swafford and Mr. Terrance E. Eccles.

Finally, we would especially like to express our appreciation to the senior executives of Tenneco Inc. for providing such a stimulating environment in which to practice our most interesting profession, the application of advanced control technologies.

T. H. Tsai
J. W. Lane
C. S. Lin

Contents

Modern Control Techniques for the Processing Industries

PART I

Process Control Fundamentals

1

Introduction to Process Control

1.1 THE CHALLENGE AND OPPORTUNITY FOR PROCESS CONTROL

The fast pace of progress in technology has resulted in making many existing plants obsolete or close to obsolete. For some plants, the only answer could be to shut the door and build a replacement using a more advanced process. This is the case when the feedstock to a plant is more expensive than the products it makes because of technological changes. For some other plants, the newer process can be just an improvement of operating efficiency from an older process, usually in terms of more consistant quality, increased yield and lower operating costs. The classical method for accomplishing this is usually by "economy of scale," i.e., to build an even bigger plant. A 200,000-barrels-per-day petroleum oil refinery is usually more efficient than a 100,000-barrels-per-day refinery with similar processing units, because more heat recovery units can be economically justified and a larger processing unit only requires the same number of direct operators as a smaller one. For this type of plant the modernization of the control system by utilizing advanced measurement and computing devices can make an old plant competitive with a new plant. This has been proven through many examples during the last few years. This type of project usually demands a much lower capital outlay than the construction of a new plant, and can offer excellent return on investment. However, the application of new technologies requires more and higher levels of managerial and engineering expertise. The application of advanced process control systems can be a very rewarding experience for engineers.

1.2 EVOLUTION OF PROCESS CONTROL SYSTEM

In the best of all possible worlds, chemical products would be made from raw materials of known and unvarying properties in equipment of unchanging efficiency to meet a fixed market in which the quantities required and the prices paid would remain constant with time. In the real world, however, conditions are far from ideal. The chemical and physical characteristics of raw materials, as well as the amounts available, change from time to time; the performance of heat exchangers, reactors, compressors, and other equipment varies with time, with the values of certain process variables, and also with ambient conditions such as air temperature, humidity, and wind velocity; and customers have a habit of cancelling or increasing their orders, or asking for better or different products.

Without changes of this kind, no control system would be necessary. In continuous processes, designs could be optimized and flows could be fixed for all time at rates which would lead to the manufacture of the required products in the desired amounts at minimum cost. In batch processes, cam-actuated valves (which pay no attention to the progress of the operation) could be used to enforce a preplanned variation of process conditions. However, since changes do occur, some form of control is required. This need for control has led to the evolution of process control technology over the past fifty years with increasing degrees of control capability. Unfortunately, most plants today represent a montage of control technologies with vestiges remaining of all previous technologies at any given time.

The evolutionary process of developing control technologies passed through three fundamental epochs—manual control, analog control, and, finally, digital control. Let's examine each of these stages in the development of the control system in terms of the definition of control. In the most general sense, "control is the action of responding to information."

In the early days control was accomplished manually by the spectator who observed the process condition or occurrences and performed a manual action which may have been either a corrective adjustment on a variable or a change of state action. Controlling the process was much like driving a car, requiring constant observation of the conditions and very frequent corrective actions in response to changes in conditions. If a significant change was overlooked due to operator inattention then the consequences were undesirable if not disastrous.

The next period saw the introduction of control devices which compensated automatically for some of the disturbances. These were called analog controllers. Analog control was accomplished by using a device which generates a signal, pneumatic or electrical, which is proportional to the value of the variable and *continuously* initiates a control action which is a function of the error of the variable signal and a reference value (setpoint). The operator determines the reference value. The value of the control output can be derived as a function of the error between the variable signal and the reference value, but also as a function of the rate of change of that error and the value of that error integrated over time.

The latest evolution in the advancement of control devices has been the application of digital computers. As in the application of any new technology, early digital computers for process control were not widely accepted by industrial users, mainly because of the high cost of computer systems and the lack of technical expertise. The growth in electronic technologies has eliminated the cost concern, and the availability of experienced engineers in computer control systems has helped place digital devices in the main stream of control system designs.

To illustrate how the new technologies have changed the scope of process control systems up to this time and why an advanced control system can drastically improve a plants performance, it is very helpful to review how and why the current control systems have been developed.

1.2.1 Elements of a Control Loop

The basic concept for automating a plant, at least by the conventional control system, is to hold several process variables at predetermined setpoints, which are usually entered by the operator. This function is accomplished by the implementation of a control loop.

Figure 1.1 shows a typical control loop, which consists of a measuring device, a controlling device, and a manipulating device. The most commonly used control loop is the regulation of fluid flow rate. The measuring device is usually an orifice and a differential pressure transmitter, the controlling device is usually a proportional integral controller, and the manipulating device is usually a pneumatically operated control valve. Note that these three devices do not have to be separate hardware components. A self-operated flow regulator is a control loop, but its application is simple and limited, and will not be discussed in this book.

1.2.2 Early Distributed Control Systems

Early automatic control systems, those implemented before the 1940s, were designed to locate the controllers close to measuring devices,

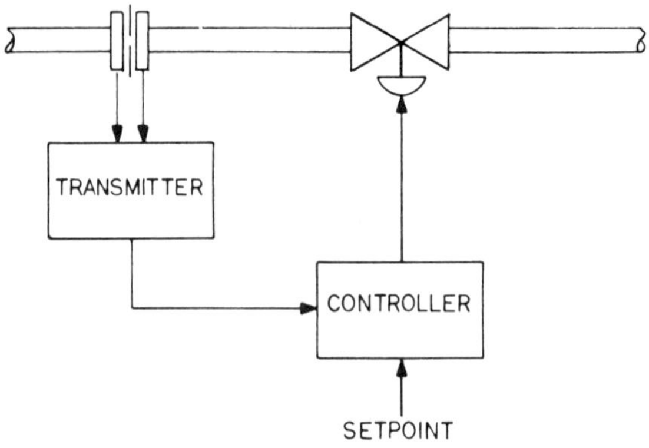

FIGURE 1.1 The four major components of a flow control loop are the orifice, differential pressure transmitter, controller, and control valve. The operator adjusted setpoint determines the flow rate.

thus the controllers were distributed throughout the plant. While this type of control system costs less than others, operators must walk around the plant to collect information and to set target values (setpoints). It was soon discovered by engineers that it was much more desirable to locate several controllers, plus other monitoring and manipulating devices, on a panel.

Figure 1.2 shows a control panel which was designed to control a boiler. Note that this panel is located very close to the process, not in a remote "central control room" as were some later day control systems. On this panel, there are controllers (with setpoint stations), recorders, indicators, annunciators, push buttons, switches, etc. Operators use these panel instruments to monitor and control this boiler. Other panels are located throughout the plant for other processes. When an operator is assigned to control a few simple processes, he uses several panels for gathering information and for entering setpoints.

All of the early control loops were based on pneumatic technology, which utilizes compressed air as a power supply. The pneumatic transmitter converts a primary signal, such as the pressure differential across an orifice, to an air pressure signal which is calibrated to be within the range of a full scale, such as 3- to 15-psi corresponding to 0- to 100-percent. The controller, which is essentially an analog computer, performs mathematical calculations by balancing air pressure with pre-set mechanical devices, and the result of the calculation is transmitted to the final manipulating device as a percent of full scale (air pressure).

The pneumatic controller is designed to process one measurement (process variable), one setpoint, and gives one result based on a predefined algorithm. Sophisticated control calculations are accomplished by linking two or more controllers together using different algorithms. Typical algorithms are proportional plus reset, proportional plus reset plus derivative, and other mathematical operations such as summation, multiplication, square rooting, etc.

Empirical and analytical methods were soon developed to compute the tuning parameters for the controllers. Ziegler and Nichols (1), and Caldwell (2) are the best known methods which were developed during the 1940s and 1950s and are still being used today.

Pneumatic controllers have one major advantage over the next generation electronic controllers; they did not present an explosive or fire hazard. Pneumatic controllers have one major disadvantage; a limitation in practical transmission distance. Thus, control panels could not be located too far away from the process.

FIGURE 1.2 Local control panels were situated adjacent to the process equipment being controlled.

1.2.3 Centralized Control Systems

The economical growth of industry after the World War II made the large scale integrated process plant feasible. To effectively monitor and control these large plants, it was desirable to improve the communication of information between processing units, thus the concept of using a large, centralized control room to control an entire plant came into being. Electronic control systems, which were first developed for military usage, became available for industrial application starting in the early 1950s. The control algorithms and design concepts used by these early electronic controllers were practically identical to those used by the previous pneumatic controllers. However, electronic systems did have advantages over pneumatic systems—the control signals could be transmitted over much greater distances and the instruments themselves were much smaller, requiring much less panel space. Thus electronic analog instruments began to dominate the market where they were utilized in large centralized control rooms. The potential fire hazard of electronic instruments was minimized using the air-purged, positive pressure control room. Intrinsically safe designs and barriers were developed later.

Pneumatic controllers were modified to counter the electronic challenge. Their size was reduced to that of electronic controllers, and practical signal transmission distance was increased by improved installation techniques and tubing design. Indeed, many large centralized control rooms continue to be designed based upon pneumatic control instrumentation.

More sophisticated control applications were developed to improve control performance using both electronic or pneumatic controllers. Typical examples are feedforward control, nonlinear control, and multivariable control. These functions were accomplished by linking a number of simple analog devices, such as proportional-integral controllers and multipliers, together. Special instrumentation expertise was needed to design and maintain these systems because of the limitations of the basic devices made it extremely difficult for process engineers to work on them. Shinskey (3) summarized these type of applications very well in his book published in 1967.

1.2.4 Supervisory Control Computers

General purpose electronic computers became available for industrial applications in the 1950s. Early applications for these computers were limited to data processing (such as payroll systems) and engineering calculations (such as solving numerical analysis problems). Some of these computers were used in the so-called batch mode, which means programs were loaded and processed by operator control. Some others were used in the so-called on-line real-time mode, which means the prestored programs are loaded and processed by time or outside events. The concept of utilizing on-line real-time computers for controlling a process plant was developed during the later 1950s.

Computer control is the technology where a digital computer is installed as an integral part of the process instrumentation and enhances the operator's ability to control his process. In order to accomplish this, the computer is connected directly to the process measurements (sensors) and other types of indicating devices which indicate the state of the process operation. The computer uses the measured data to determine the current state of the process. Having found the current state, this information is evaluated using a model of the process which is being controlled to determine where the process could be operating to achieve the most profitable operation. These calculated conditions are checked for feasibility, and if found *not* to violate any operating constraints, are transmitted back to the process via a set of control equations. The computer is capable of performing these complex calculations, evaluating the alternatives, selecting a control strategy, and implementing control actions in the matter of seconds. Thus, the computer is essentially performing real-time control, that is, it is evaluating the process operations and taking control actions within a meaningful time span.

When process control systems were first installed, they were interfaced with the existing conventional analog instrumentation whether it was pneumatic or electronic. The early systems were not powerful enough or fast enough to perform the regulation function, therefore this function remained with the analog controller. The computer generated setpoint changes to these controllers. Figure 1.3 shows a typical block diagram of the early process control system configuration.

Because of the high cost of implementing a process control system a candidate had to be very carefully selected by a comprehensive feasible study. Note that a supervisory computer is not necessary for plant operation. It can only be justified when an identifiable benefit can be generated by the installation.

The major advantages for using a supervisory computer are that all process measurement data are available from one common device, and that sophisticated calculations can be programmed and executed at a desired time and frequency. This makes the application of the multivariable control algorithm feasible.

Modern control theory (4) was introduced for industrial application. This type of control theory is based on sample data, time domain analysis, as compared to the continuous, frequency domain analysis which is the basis for conventional control theories. In addition, multivariable controller design can be easily formulated.

A side benefit from the application of supervisory control is the ability to save large amounts of operating data for engineering analysis,

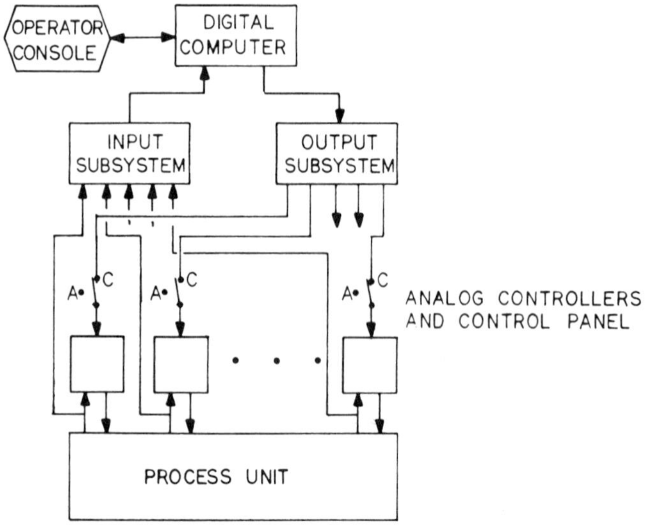

FIGURE 1.3 Supervisory computer system interfaced with the process via existing analog controllers and computer input/output subsystems. Operator console provided communications interface between the operator and the process control system.

that is, to use the computer as a data logger. Indeed some supervisory computer applications were partially justified by data collection and reduction.

The application concept for a supervisory computer has changed considerably over the years. As engineers became more familiar with the capabilities of computers, more and better applications were possible.

1.2.5 Direct Digital Control Systems

The term Direct Digital Control (DDC) was first introduced in the early 1960s (5). The basic concept of DDC is to use a digital computer to replace and perform the function of analog controllers. Since computers were expensive in the 1960s, it was necessary that the DDC computer be capable of replacing several hundred analog controllers to be economically competitive. Figure 1.4 shows a DDC system configuration.

There were strong arguments for the use of DDC computers, such as:

More precise controller settings
No controller drift
Centralized display and tuning of controller parameters
Ease for control configuration modification
Ease to interface to a supervisory computer
The use of advanced control algorithms
The use of adaptive controllers

While most of these arguments for DDC were valid, there existed a very serious shortcoming for centralizing too many controllers in one common device, that was the consequences of an equipment failure. A DDC computer failure created chaos for operators because too many control loops fail at the same time. This was the main reason for the poor acceptance of early DDC computers. To counteract this deficiency, back-up analog controllers or even back-up computers were installed, but this increased the installation cost making it unattractive in most instances.

There were other shortcomings for the early DDC computers. One was the limitations of the operator communication devices, and another one was the difficulties encountered in software maintenance. Furthermore, although advanced control algorithms could be implemented by DDC computers, most DDC loops simply used the digital form of the Proportional-Integral-Derivative (PID) control algorithm. The performance of a DDC system, even when it was running, was not much more superior than an analog control system.

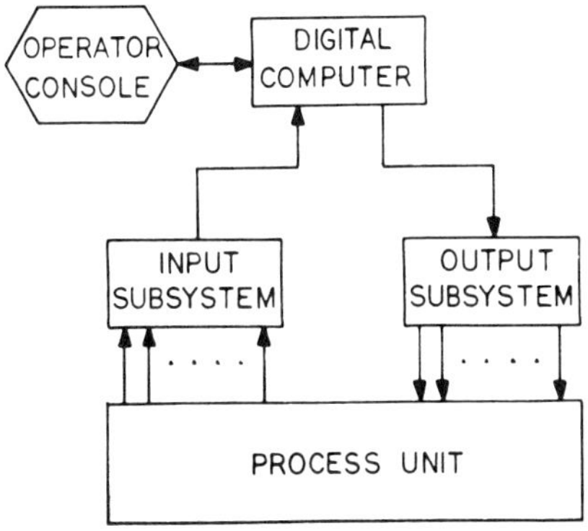

FIGURE 1.4 Direct digital control system interfaced directly with the process measurement signals and control devices without requiring analog controllers. Operator console provides communications interface between operator and the process control system.

There were many misapplications of DDC systems (and supervisory systems) during the 1960s. In fact, several large DDC computers were removed after a few years of operation and replaced by analog control systems. These particular applications have to be considered as technical and economical failures. However, the experience and knowledge gained by these applications provided the incentive to develop the highly successful distributed control systems of recent years.

1.2.6 Modern Distributed and Integrated Control Systems

The availability of low cost microcomputers and communication devices in the 1970s made it possible for instrumentation and control companies to develop distributed digital control systems. The first distributed digital control system was announced in the mid-1970s by one vendor, and many other new systems have been introduced into the market since then. Distributed digital control systems are now applied in most new process plants.

Figure 1.5 shows a typical distributed digital control system. There are four major components to make a complete system.

1.2.6.1 Digital Controllers

Microcomputers are the basic building blocks for the digital controller. The design concept is basically using a microcomputer to functionally replace one to several conventional controllers, computing devices, switching devices, etc. The control algorithms are generally

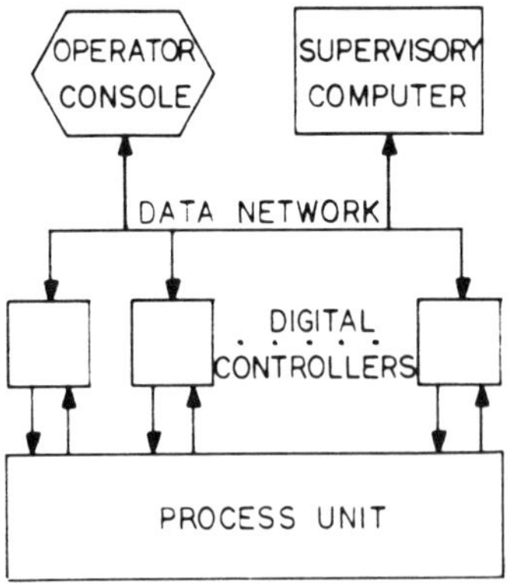

FIGURE 1.5 Distributed control system consists of microprocessor based digital controllers distributed throughout the process, a digital communications data network, a central CRT based operators console, and a supervisory control computer.

preprogrammed in linkable blocks and users may configure a control loop by linking these computing blocks together to accomplish the desired strategy.

1.2.6.2 Operator's Consoles

The operator's computer console is used to replace conventional control panels. A console usually consists of one or more cathode ray-tubes (CRTS) with keyboard, a printer for messages, and may also include special annunciation panels and chart recorders. Some consoles also include a dedicated computer for data storage, display generation, and other functions.

1.2.6.3 Data Links

Serial data communication channels, usually redundant, are used to transfer information between digital controllers and the operator's console. Usually these channels are designed to facilitate many controllers.

1.2.6.4 Supervisory Computer

The supervisory computer is usually considered by users as optional equipment because it is not necessary for accomplishing basic control of a plant. A supervisory computer can be easily interfaced to the distributed digital control system using the same communications channel that connects the operator's console to the digital controllers. The cost for this type of connection is much less than that required to interface with a conventional instrumentation system as the one shown in Fig. 1.3. There are also many standard supervisory control software packages which reduce programming costs. Most distributed digital control systems easily accommodate a supervisory computer.

There are many advantages for using a distributed digital control system as compared to using a conventional analog control system. These include lower cost, superior performance, greater reliability, and increased flexibility. Distributed control systems have been installed in practically all new process plants built since 1980. Many existing plants with conventional control systems also have been retrofitted with distributed control instrumentation.

The design concept for implementing distributed control systems for a new process plant has mostly followed that for implementing analog control systems, except for some sophisticated users. This is because a proven technology in process and control system designs is represented by analog control systems. Since a great majority, if not all, of the new plants are built by using proven technologies, distributed control systems are being implemented by a loop-for-loop replacement approach based on existing piping and instrumentation diagrams with the exception for the design of control panels which are replaced by CRT displays. This concept is a practical and easy way for implementing distributed control systems, and can also be considered the right way because one cannot develop a new control concept while the plant is being built. However, a distributed control system with a supervisory computer can be used to implement advanced control functions which will generate significant profit. It is extremely important for plant operation that these advanced control functions are identified and developed after the plant is built. A distributed control system for a new plant, or for replacing a large existing plant's control system, should be designed to provide a means for easy implementation of advanced control functions at a later date. Otherwise, the cost for adding advanced control functions could be excessively expensive, and they may never be implemented.

Note that the expertise required for the design and implementation of conventional control systems resides mostly with instrument engineers, who are not necessarily experts in designing advanced control strategies. Process engineers with working knowledge in instrumentation, computers, and plant operations are needed for this task. The next section explains the concept of controlling the performance of a plant and the basic approaches to accomplishing this end.

The concept of the "integrated control system" has been developed since the mid-1980s, and this type of system is evolving at a very fast pace. The fundamental purpose of an integrated system is the sharing of a common data base by all concerned departments within a plant, so that all information and control systems can be integrated together. A good example of such a system with a limited scope is the integration of a laboratory analysis computer and a supervisory computer permitting the test results from laboratory analyses to be used to check the on-line analyzers or to control product quality.

There are many different architectures for integrated systems, but the major technologies that make this type of system desirable are the availability of high speed digital communication systems and low-cost, high-performance microcomputers.

A well-designed integrated system offers two major advantages over the relatively older distributed systems. They are the linking of different computers using standard "gateways," and using a multiplicity of computers to replace a single supervisory computer. The so-called "gateway" is usually a microcomputer especially designed to interface two different types of computers, so users do not need to develop special communications software. The purpose of distributing the central supervisory computer functions to several special purpose computers is to simplify the software design and to improve system reliability, flexibility, and speed.

1.3 THE APPLICATION OF ADVANCED CONTROL SYSTEMS

1.3.1 Design and Operating Criteria

The basic goals for operating a plant are simple and straightforward. They are:

Maximize safety
Adhere to governmental regulations, such as environmental standards
Maximize or control on-specification product production rate
Minimize operating cost

These criteria are also shared by design engineers. For most processes a plant cannot be designed to stay at an optimum condition. In real life "optimum" is an elusive target which is ever changing. Consequently, a process must be designed to provide stable operation over a reasonable operating range of throughput and feedstock characteristics, and all equipment must be sized for the worst case. Without the help of an advanced control system, operators usually set the process at the safe side of the operating range, or the condition least likely to cause plant upsets when disturbances enter the process. This usually means keeping the process at an operating condition which requires higher operating costs than that at the optimum condition. This concept can be best illustrated by the design and operation of a distillation column.

Figure 1.6 shows a simplified process control diagram for a deisobutanizer, which is designed to separate a mixture of isobutane and normal butane into the two relatively pure products. Since the boiling points of these two components are close, a large number of trays must be used in the column and a fairly high reflux-to-feed ratio is required to maintain the product qualities within the specifications. Higher reflux ratio increases product purity, stabilizes the process, but also increases the energy cost.

Figure 1.7 shows an operating curve for a deisobutanizer based on Griffin's (6) test data. Note that although the specifications for isobutane allows a normal butane content of 4.0 percent, manual operation of the plant produces much purer products than the specifications call for, and this results in more energy being expended than necessary. This is because of the operator's desire to keep the process at the safe side, which can be seen by comparing the risk in product quality under manual operation with that under otpimum operation as shown in Fig. 1.7. Note that under manual operation, a 10 percent reduction (from 8 to 7.2) in reflux-to-feed ratio causes a small increase

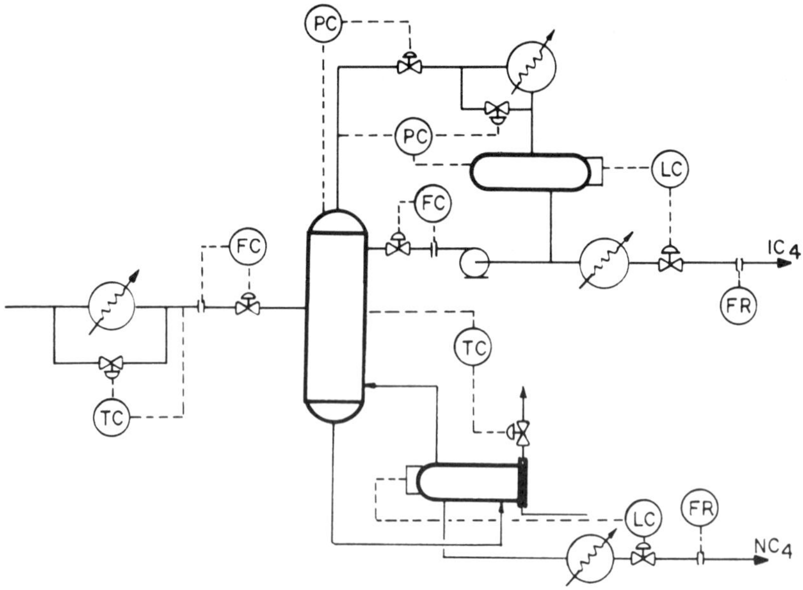

FIGURE 1.6 Simplified process control diagram for deisobutanizer distillation column showing typical control instrumentation for column, overhead accumulator, and bottoms reboiler.

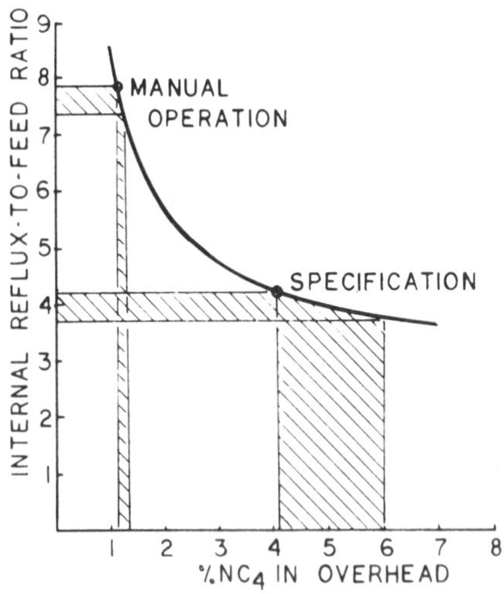

FIGURE 1.7 Deisobutanizer overhead product operating curve showing relationship between internal reflux-to-feed ratio and normal butane concentration in the overhead product. A given variation in ratio at high ratio value produces a much less variation in product composition than the same variation in ratio at a lower value.

in product impurity (from 1.1 to 1.35 percent), while at the optimum condition, the same 10 percent reduction (from 4.2 to 3.78) causes a much greater increase in product impurity (from 4 to 5.6 percent). Note also that any variation in product quality within the specification does not concern operators, while any variation exceeding the specification concerns operators because they must make adjustment in the process to bring the product quality within the specification.

Since operators can not observe the quality of one product all of the time, it is a logical decision for him to keep the process at a less optimum but more stable condition so the product quality will consistently be better than the specification. The real challenge for the control of a chemical plant, such as a deisobutanizer, is to meet all specifications and constraints at the same time while keeping the process at the optimum condiition.

The most energy efficient condition for operating a deisobutanizer is to reduce the reflux ratio and heat input to the reboiler to a state that both the overhead product (isobutane) and the bottom product (normal butane) are at the specified purity. This will also be the most economical condition when the price differential between normal butane and isobutane is small. To maintain the process at this condition would be a very difficult task unless a sophisticated control system is installed.

Of course every process has its own special characteristics, but the basic goals for control are about the same and process performances will be greatly affected by how they are controlled.

1.3.2 Process Engineers and Control Analysis

A process engineer's responsibility in an existing process plant is to maintain and to improve the plant's performance. The major tasks are to identify and remove process bottlenecks which limit the plant's performance in terms of throughput, yield, energy consumption, etc. A design-oriented process engineer tends to look at the process at steady state conditions, and the classical approach to investigate the unit performance is to identify the bottleneck equipment and replace it. While this approach is needed when a plant is first built, efforts like these soon reach a point of diminishing return when the equipment in a plant is well balanced. This is the condition when engineers realize that the bottlenecks change from time to time. These type of bottlenecks cannot be removed by replacing equipment, only an improvement in the control system will help.

To identify changing bottlenecks is a challenging task. Indeed, a process plant can be so complicated that a complete understanding of how to operate it is elusive. It is rather common to observe that different operators have very different understandings of a plant's operations and consequently, different philosophies on how to run the same plant to achieve similar performance levels. This is obviously a system which has more variables than equations, thus an infinite combination of valid operating conditions exists. In addition, a process plant is practically always in a transient condition. The so-called steady state operating condition is simply a transient condition which varies within a small range. This makes the collection and analysis of operating data a difficult and interesting task, which is also one of the most important tasks for a process control engineer.

The final goal for collecting and analyzing operating data, of course, is to identify the cause-effect relationships between process variables, or to define mathematical models for the plant. Note that the so-called steady state models are time independent, thus they are in the form of algebraic equations; the so-called transient or dynamic models are time dependent, thus they are in the form of differential equations. Techniques for developing process models will be reviewed in detail elsewhere. The following paragraphs discuss the basic concepts for developing models.

1.3.2.1 Apply Statistical Correlation with Extra Care

Because of the large number of process variables involved in process identification, one method favored by some engineers has been the use of statistical correlations to determine the cause-effect relationships between process variables. This can be seen by the availability and application of multivariable regression analysis packages. While the statistical correlation procedure is an acceptable method to determine the mathematical relationships between variables, the blind application of this technique without considering the underlying chemical and/or physical relationships of the process can result in very misleading conclusions at times. This can be seen by considering the following example.

Figure 1.8 shows a simplified process control diagram for a crude oil furnace. Oil inside the pipe is heated in the furnace to a desired temperature, which is normally controlled by manipulating the fuel flow rate. A temperature controller is generally used for this purpose.

Should an engineer try to develop a furnace model by correlating raw operating data taken while the temperature was under closed-loop control, he would find the set of data shown in Fig. 1.9, which shows that the oil temperature decreases with increasing fuel flow rate.

Of course oil temperature increases with increasing fuel flow, that is, when there is no disturbances to the process. However, when the oil temperature is under closed-loop control, and the fuel heating value drops because of other process upsets, the oil temperature would appear to be decreasing while the fuel flow is increasing as shown in Fig. 1.9. A regression analysis based on these data without considering the fuel heating value would result in a process model whose apparent gain is reversed from the real process.

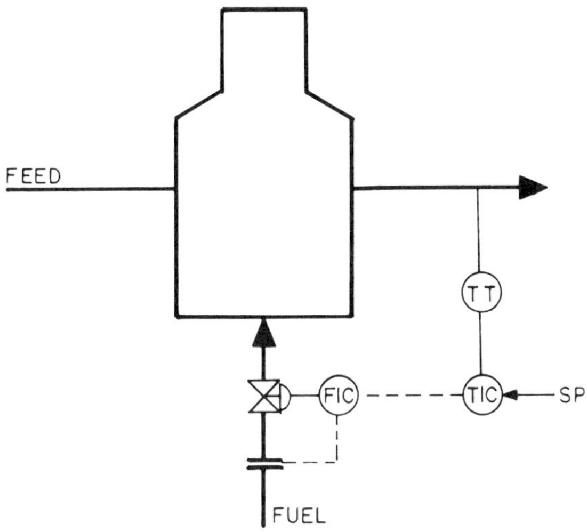

FIGURE 1.8 Simplified product temperature control diagram for a furnace. Temperature controller is cascaded to fuel flow controller.

Most engineers will not make this mistake and conclude that the oil temperature reduces with increasing fuel flow, regardless of whatever the regression analysis implies, because a furnace is a simple and fast process. The increasing fuel flow rate being the result of closed-loop control of temperature is obvious. However, for a more complicated,

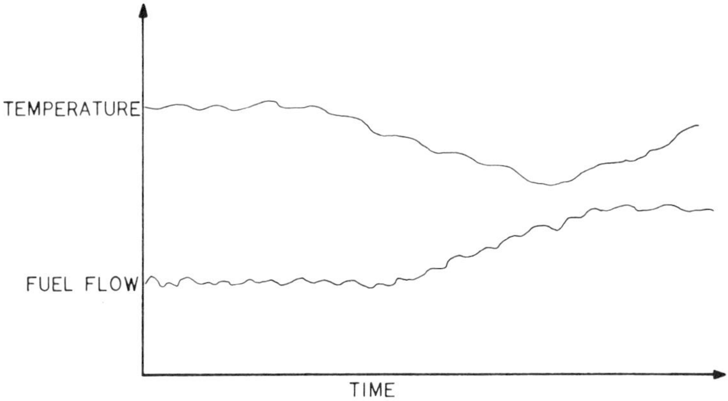

FIGURE 1.9 Temperature response to decreasing fuel heating value when temperature is controlled by fuel oil flow. The fuel oil flow correction lags the temperature deviation.

slow process where the feedback loop control is performed manually, a mistake can be easily made. The following mistake was actually made by a well-educated engineer.

The process is a large paper machine producing heavy board. The basic specifications for the product are basis weight (measured by pounds per 1000 square feet), moisture (measured by weight percent of water in the product), and caliper (thickness measured by thousands of an inch), plus several others. The bottleneck for throughput when making heavy board is usually the dryer capacity, that is, an increase in machine speed will result in an increase in product moisture. As in operating any other process, operators usually maintain a safety margin for machine speed so the product can meet moisture specification when disturbances enter the process, thus keeping the machine speed somewhat below the maximum. This well-educated, but less experienced engineer was assigned to evaluate how to maximize the speed.

After weeks of data collection and computer simulations, this engineer concluded that in order to maximize the paper machine speed, operators should set the moisture at the minimum, that is zero percent, because statistically, the paper machines run faster when the moisture is lower. The confidence level of his correlation was very good.

All the production people who had real experience on paper machines knew this conclusion was wrong, but could not tell where the mistake was made in a large set of mathematical equations and data. A more careful analysis by a more experienced engineer found out that over a long period of time, lower machine speed coincides with higher moisture. However, this is caused by the variations in pulp quality and the degree of refining of the pulp. It was well known that when the pulp contains longer fibers, more water can be removed at the wet end section, thus less water needs to be removed in the dryer section, so the operators can increase the machine speed. Since operators want to keep a safety margin, they can increase the machine speed to a fast enough level by their standard, but still below the dryers' capacity, so the paper is dryer than normal at higher machine speeds. (This condition happens only when the moisture is not under closed-loop control by manipulating the steam pressure to the dryer). To conclude that one should try to make the paper as dry as possible to increase machines speed is just like saying that the furnace temperature increases with reducing fuel flows, but this conclusion was reached as a result of improper statistical analysis.

Another important reason for the need for special care in applying statistical analysis of large amounts of data is because of the GIGO (garbage in-garbage out) limitation. Most plants' measurement devices can give a reasonable repeatability on the measured variables, but they are not usually accurate. When large amounts of data over a long period of time are used for process identification, the lack of accuracy could easily cause very poor results.

1.3.2.2 Break the Title Barrier

Most engineers in a plant are assigned a title such as "process engineer," "instrument engineer," "systems engineer," etc. Many people become specialized in one field because of the job assignments for that title. A process engineer can be an expert in material and heat balance calculation, thermodynamics, kinetics, fluid flow, etc., but knows very little about instrumentation, control system design, or computer systems. An instrument engineer, or a system engineer, can also be so specialized in their own field that they depend on process engineers for process information. Specialization in one field may be desirable for a member in a large project team which divides the tasks into many areas and assigns special people to define and integrate the tasks, but is not desirable for an engineer whose job is to analyze and to improve the performance of an existing plant. An engineer needs broad knowledge in many fields in order to develop effective answers. The combination of several specialists may not match a single person who has a broad knowledge in many fields. An engineer does not speak a computer programmer's language and visa versa. An engineer without any computer knowledge will not be able to design effective control systems using a computer by depending on a programmer to write the computer specification.

For example, when an engineer designs a sample data control loop at a 10 second sampling interval, he expects the time interval to be exactly ten seconds, a small error may be acceptable, but certainly not an error exceeding ±0.5 seconds under normal conditions. A programmer, unless he understands sample data control, may not appreciate this need for precise timing at all. It is acceptable to a programmer when the computer processes this loop 1000 times in 10,000 seconds, even when the intervals can be between 5 to 15 seconds. This condition did happen in the implementation of a specially designed algorithm, and delayed the implementation by several weeks because the engineer who designed the algorithm did not realize the errors in computer timing and thus made unnecessary modifications to his algorithm design.

1.3.2.3 Be Objective

It is important for an engineer to remember that the only goal in implementing an advanced control system is to improve the plant's performance. Many "nice to have" functions may not be all that nice when these functions are competing for computer time with important control functions. Some years ago a company specified a supervisory control computer with 12 typers and a line printer to print out the "nice to have" reports. They later realized that it not only cost a lot of money to purchase these typewriters, it was detrimental to the control system performance. To make the matter worse, no one wanted to look at these reports anyway. So these typewriters were eliminated and the performance of this system improved.

Another important rule to follow is to try the simplest method possible. Many engineers, particularly well-educated ones, tend to over specify many problems. When a simple feedback controller can do a satisfactory job, there will be no need for a complicated process model.

1.3.3 Process and Control System Design

During the years when single-loop controllers, relay based interlocking systems, and hand-wired panel boards were state-of-the-art, process and control system designs were generally delegated to two separate groups. The basic criteria was to design a plant stable enough for manual control. The so-called fully automated plants, at the time, consisted of more control loops and longer instrument panels than others, but the real control actions, such as the selection of optimum setpoints and batch sequencing functions, were done by operators. This criteria has been so deeply rooted in the minds of so many engineers that even today, the basic design concept for control systems has changed little. Most distributed control systems are still installed to replace existing analog control systems on a loop-by-loop basis, and most Programmable Logic Controllers (PLCs) are still used to replace existing relay based interlocking systems. Furthermore, a continuous process is still considered as superior to a batch process because a batch process "cannot" be automated. This implies that most of the advanced capabilities offered by an up-to-date control system are not being used or not even being recognized by engineers who are responsible for process development. This also implies that the opportunities for the application of advanced control systems are wide open, and engineers who can combine process development and control system development together can make great contributions to the process industries.

REFERENCES

1. Ziegler, J. G. and N. B. Nichols, Optimum settings for automatic controllers, *Trans ASME, 64*:759–768 (1942).

2. Caldwell, W. I., Frequency response analysis, *Taylor Tech, 4*(2): 13–19 (1951).

3. Shinskey, F. G., *Process Control Systems*, McGraw-Hill, New York, 1965.

4. Tou, J. T., *Modern Control Theory*, McGraw-Hill, New York, 1965

5. Williams, T. J., Direct Digital Control Computers—A coming revolution in process control, *Proceedings Texas A & M 19th*

Annual symposium on instrumentation for the process industries, College Station, Texas, 1964, pp. 70–81.

6. Griffin, D. E., Fractionator controls can save energy, *Oil and Gas Jour.*, March, 1981. pp. 140–154.

7. Tsai, T. H. and J. W. Lane, Experiences in Batch Chemical Process Control, Computer Control of Batch Processes, AIChE Workshop, May, 1976, pp. 15–21.

2

Process Measurement Techniques

2.1 INTRODUCTION

Devices that provide accurate measurement of process variables are essential for controlling the process in actual practice. These provide the operator the window into internal conditions of the process he requires in order to evaluate the performance of the process. These are also the fundamental elements for any automatic control device provided to regulate a particular process variable.

Supplementing the fundamental devices for measuring flow, temperature, and level are a large variety of analyzers which measure the physical and chemical properties of process streams.

There have been numerous technical papers presented and articles published on field process measurement techniques (1). The purpose of this chapter is not to duplicate these works, but to emphasize the important concepts the process engineer should be familiar with in order to effectively utilize these measurement devices in the design of a control system or to evaluate the performance of a process.

2.2 ON-LINE CONTINUOUS MEASUREMENT SYSTEMS

The basic design concept for providing a continuous measurement is converting a physical property into a continuous signal which is related in magnitude to the physical property being measured. This signal is then transmitted in some form, usually mechanically, pneumatically, hydraulically, or electrically, to a remote location to be used for display or control.

Figure 2.1 shows a typical flow measurement unit. Note that there is a sensor (the orifice plate), a transmitter (differential pressure cell), and an indicator. In addition, there are other supporting devices such as power supply, panel and housing, wires or tubes, to complete a measurement point. Similar types of devices are also used for measuring other variables, such as temperatures and pressures. For the purpose of discussion, a flow measurement device will be used as an example.

2.2.1 Sensors

A sensor is the element which responds to the physical property one wishes to measure, translating the value of the physical property into a corresponding signal. The ideal sensor is a device which only responds to one property and nothing else. This is an ideal condition that is not achievable in most cases, so a sensor is usually designed so that the effects of other physical properties are small or can be compensated for by other measurements.

Typically, for each physical property to be measured, there are many types of sensors (2,3). Some of them are designed for special applications, but many of them are designed to compete for identical applications. For example, to measure liquid flow rate through a pipe, one can select from several alternatives including positive displacement meters, differential pressure meters, ultrasonic meters, electromagnetic meters, vortex shedding meters, and a few others. Each type is also available in many different designs. It is not possible, nor desirable, for a process control engineer to become familiar with every detail about all sensors, but it is important that he understand how a sensor works, and how accurate the results can be. The accuracy specified

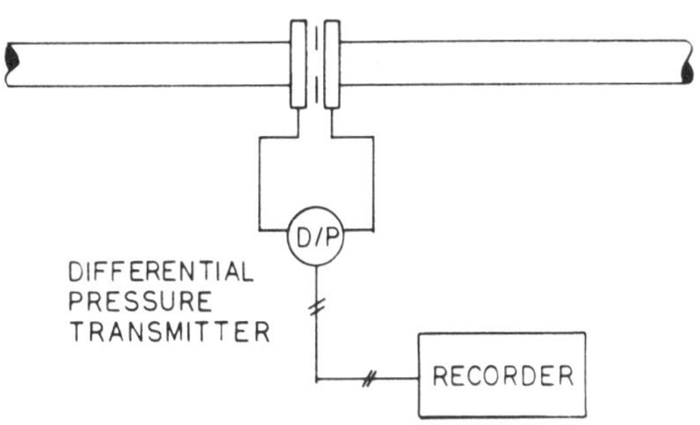

FIGURE 2.1 Flow measurement system including orifice plate, differential pressure transmitter, and signal recorder.

by a sensor's supplier may be based on a predefined set of conditions, and it is an engineer's job to determine how much confidence he can have in the performance of a specific sensor based on the actual operating conditions.

Let's take the most commonly used orifice meter as an example. Figure 2.2 shows an orifice meter. As shown in many text books (4), the volumetric flow rate at the base conditions can be computed as:

$$F = \frac{C_o \times f_{Re} D_o Y}{\rho_b} \sqrt{2 g_c (p_a - p_b) \rho_a} \tag{2.1}$$

where:

F = Flow rate at base condition

C_o = Basic orifice coefficient

f_{Re} = Reynolds no. correction factor (equal to 1 when Re. no. is very large)

D_o = Diameter of orifice

Y = Compressibility correction factor (equal to 1 for noncompressible fluid)

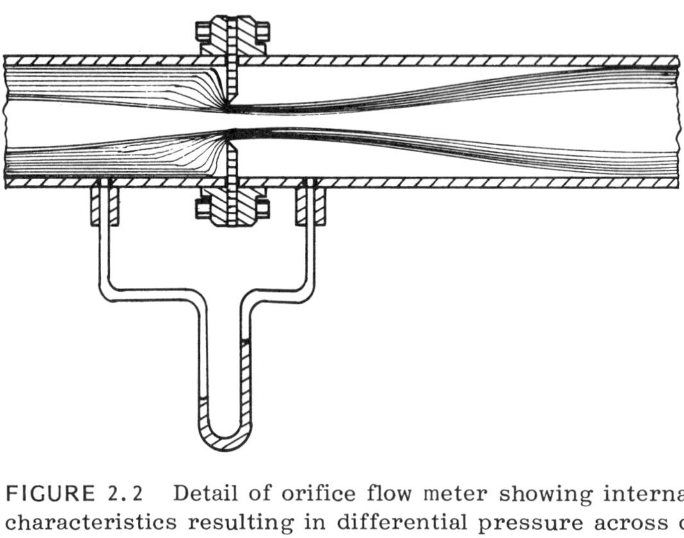

FIGURE 2.2 Detail of orifice flow meter showing internal fluid flow characteristics resulting in differential pressure across orifice plate. Differential pressure is indicated by level difference of fluid in u-tube manometer.

ρ_b = Density of fluid at base condition

ρ_a = Density of fluid at flowing condition

g_c = Gravitational force

$p_a - p_b$ = Pressure differential between the two pressure taps on the pipe

In actual practice, only the differential pressure is measured so the flow rate computation can be simplified to:

$$F = K \sqrt{p_a - p_b} \tag{2.2}$$

where K = meter constant for the particular orifice.

Comparing Eqs. (2.1) and (2.2), it is clear that many flow meters using only orifice plates are designed by assuming that all variables, except the differential pressure, including fluid density, are constants. Therefore, a flow meter design determined by this means is subject to errors when the operating conditions deviate from the design conditions.

To correct this potential error, in cases where the fluid is not incompressible and, thus, more subject to variations in temperature and pressure, temperature and pressure corrections are included in the flow calculation which now gives mass flow. The equation for this calculation is:

$$F = K_1 \sqrt{\frac{P_a \times (p_a - p_b)}{T_a}} \tag{2.3}$$

where:

K_1 = Meter constant

P_a = Absolute pressure at flow condition

T_a = Absolute temperature at flow condition

This so-called mass flow measurement is also subject to errors, which can be caused by composition changes, and when the gas deviates from the assumed ideal-gas law, such as when under very high pressure.

In addition to the errors caused by the changes in operating conditions, the age of the orifice plate may be another source of error. As shown in Fig. 2.2, the pressure drops to a minimum point at a distance from the orifice plate, and this distance depends on the sharpness of the orifice plate. A slight wearing of the orifice plate, or the deposition of some solid material on the plate, such as scale, will change the flow pattern. Since the measuring point is fixed, the differential measure will change at a given flow rate.

When the accuracy of a flow rate measurement is important, and the potential errors encountered from using orifice plate are unacceptable, an alternative sensor should be considered. The accuracy of positive displacement meter, for example, will not be affected by composition changes.

The errors discussed so far are all in the static error category, that is, they are not time dependent. When the measurement result is used for closed loop control, dynamic response of the instrument becomes very important. This will be discussed later in this chapter.

2.2.2 Transmitters

There are two broad categories of transmitters—pneumatic and electronic. Like the sensors, there are many different designs for transmitters.

2.2.2.1 Pneumatic Transmitters

Pneumatic transmitters use compressed air as their power supply. Figure 2.3 shows a simplified diagram for a force balance pneumatic

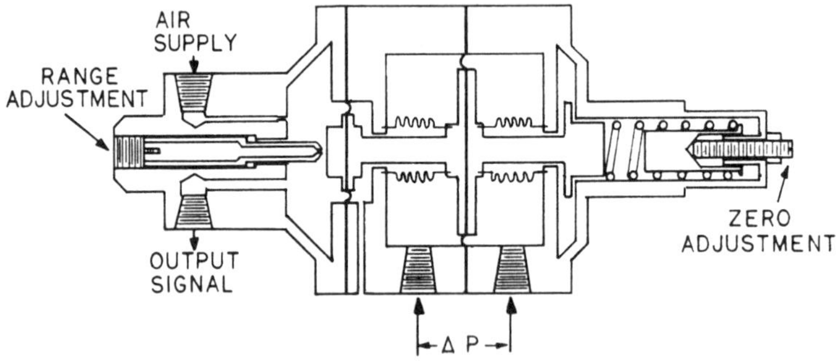

FIGURE 2.3 Internal construction of pneumatic force balance differential transmitter which converts a low differential pressure signal, ΔP, into a proportional high pressure output signal. Increasing differential pressure moves center diaphram on piston to left, thus, reducing gap between the left piston face and leak orifice. Consequently, pressure builds up in chamber increasing output signal until all forces on piston are balanced.

transmitter, which amplifies a weak differential pressure signal, measured in inches of water, to a much stronger signal, measured in pounds per square inch (3 to 15 psi). The transmitted output pressure is directly proportional to the differential pressure (or pressure, or other variables from other sensors). A pneumatic transmitter usually has two adjustments, zero and span. When properly adjusted, the 3 to 15 psi signal represents 0 to 100 percent of sensor measurement range, such as 0 to 150 inches of water.

The accuracy and repeatability of a transmitter are usually specified by the manufacturer when calibrated under well-defined conditions. Periodic recalibration of transmitters is needed to maintain their accuracy.

Pneumatic transmitters have another potential error source; that is, the pressure drop when the output signal is transmitted over a long distance. Because of the concern for pressure drop, the functional operating distance between the pneumatic transmitter and the controller is usually limited to a maximum of 350 feet, unless a booster is used.

2.2.2.2 Electronic Transmitter

Conventional electronic transmitters were first designed based on the same physical fundamentals from which pneumatic transmitters were designed with DC power replacing compressed air as power supply. For example, the most popular conventional electronic differential pressure transmitter is based on the force balance concept. The standard signal range for output from the transmitter is 4 to 20 milliamps (ma) which represent 0 to 100 percent of the measurement range. This type of transmitter also has zero and span adjustments for calibration, and the accuracy and repeatability, depending on the manufacturer, are in about the same order of magnitude as that for pneumatic transmitters. Other than the capability to transmit the output signal over long distances without pressure drops, conventional electronic transmitters do not have a clear advantage over pneumatic transmitters. However, the availability of electronic transmitters has helped promote the development of many new sensors, which are based on the measurement of electrical properties such as resistance. These new sensors have greatly improved measurement accuracies.

When calibrating a transmitter, whether pneumatic or electronic, it is adjusted to give a signal proportional to the signal of the sensor within a specified tolerance. Supposing the reading of a sensor is 50 percent, then the reading of a transmitter, with a specified accuracy of ±0.1 percent of full scale may vary from 49.9 to 50.1 percent. When a transmitter is subject to drift, the error could become greater (5). The calibration of a transmitter only assures that the signal of the sensor is converted correctly, but the sensor itself may not be part of the calibration, so the total error of a measurement may be somewhat greater than the transmitter error alone.

2.2.2.3 Smart Transmitters

The application of microcomputers has greatly improved the accuracy and reliability of electronic transmitters (6,7). The so-called "smart transmitter" has been commercially available since 1982. The brain of the smart transmitter is a microcomputer which provides such standard capabilities as linearization, self-diagnostics, and bidirectional digital communication permitting calibration, verification, and analog signal transmission over the same pair of wires. This is the beginning of a new concept of transmitter design. At this time, the cost of smart transmitters is greater than that for conventional transmitters, and they are only available for a limited range of applications. Their capabilities are mostly designed for reducing maintenance costs, which is the major justification for purchasing these more expensive transmitters. Like most other high technology products, there will be more and better products available at a lower cost in the near future, and smart transmitters could become mainstream products instead of being considered as special products. It is possible that in the near future, all process data transmission will be in digital form and the 4 to 20 ma standard signal for electronic analog instruments will become a thing of the past, while the errors due to transmitters become negligible. Note that some specialized instruments already have automatic calibration capabilities. Process control engineers should pay close attention to these developments.

2.2.3 Indicators and Recorders

Indicators and recorders are used to display the results of conventional transmitters. Some of them are used in conjunction with controllers which will be discussed later in this chapter.

The linear position-type servomechanism indicator (or recorder) is one of the devices frequently used as part of a conventional instrumentation system. This device drives a servomotor whose position output is proportional to the magnitude of an analog signal. Figure 2.4 shows a simplified electronic design for an indicator/recorder. Similar devices have also been used to display pneumatic signals. The accuracy of an indicatory/recorder depends on the quality of hardware design and the calibration, just the same as that for transmitters.

One of the most significant sources of measurement error could be the difficulty of reading the recorder chart or an indicator. This is especially true when reading nonlinear signals such as those obtained from a flow measurement orifice. The flow rate is proportional to the square root of the differential pressure signal. The most convenient method for interpreting this signal is to include a square root extractor device in the signal circuit, which will linearize the measurement signal.

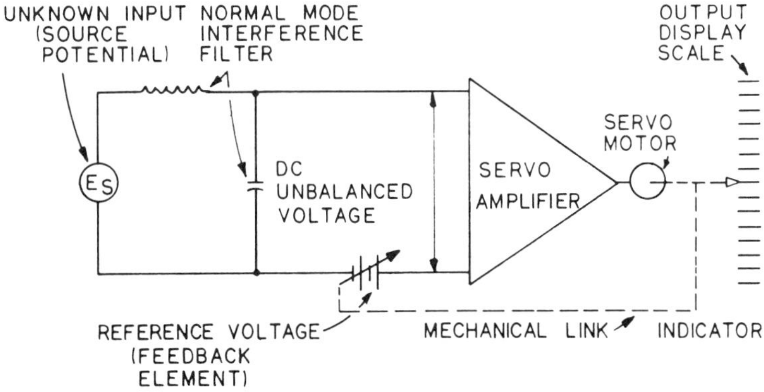

FIGURE 2.4 Circuit diagram of linear position type servomechanism which converts input signal value to indicator position on the display scale.

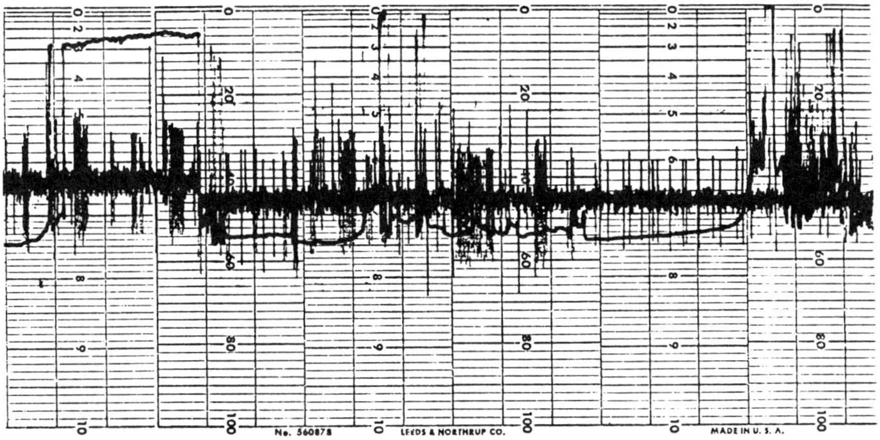

FIGURE 2.5 Typical strip chart recording of process variable printed with linear and logarithmic scales.

However in many installations, the square root extractor device is eliminated as a cost cutting step, and the nonlinear signal is displayed on a logarithmic scale either as an indicator or on a strip chart.

In fact, most general purpose recorder chart paper includes both linear and logarithmic scales as shown in Fig. 2.5. Note that when the flow rate is low, say at 30 percent full scale or less, reading errors can be very high. In addition, high frequency noise can make the readings even less accurate.

The best way to interpret the results from a chart, for process analysis purposes, is to integrate the measurements for a period of time during steady operation, and compute the rate from the integrated value if needed.

The advance of modern distributed control systems has changed the methodology of displaying process data with greatly improved accuracy. Basically, these systems use digital data and display the results, either numerically or graphically on cathode ray tubes (CRTs) or on log sheets. Data can be collected in numerical forms automatically and stored on magnetic devices (disk or tapes) for further analysis.

2.3 ANALYZERS

There are numerous physical properties which must be measured by analyzing a small sample drawn from the process stream because continuous sensors are not available. One such property is the composition of fluids. Gas chromatographs are the analytical technology most often utilized for this purpose.

Analyzers can be classified into two major functional catagories: on-line and off-line. On-line analyzers are those whose sampling lines are directly connected to process streams, so the samples are analyzed automatically without requiring operator intervention. Off-line analyzers are those where the samples are collected manually and subsequently presented to the analyzer for analysis. On-line analyzers are used for process monitoring and control when frequent analysis of process streams are important.

2.3.1 On-Line Analyzers

An on-line analyzer system consists of four parts: sampling system, sensor (or analyzer), output signal generation, either digital or analog, and recorders (or printers). This is highly specialized technology, and the technical details will not be discussed in this chapter.

From the standpoint of the system designer planning to utilize on-line analyzers, two important decisions must be made. The first, should a single-stream or a multistream sampling system be used, and the second, where are the sampling points to be located.

On-line analyzers can be expensive. One way to cut the cost is by using one analyzer for analyses of a number of samples taken from different locations. Figure 2.6 shows a simplified diagram for such a system. The analyzer is used to test three streams plus a standard sample, which is used to calibrate the analyzer. This arrangement reduces the cost of analyzers as compared to one using a dedicated analyzer per stream, but it also increases the sampling cycle time of each process stream. If the testing results are used as input for a control function, the impact from the slowdown of sampling frequency on the control performance should be analyzed. In some cases, a single-stream sampling system is more desirable in spite of the higher cost.

The location of the sampling point is a straightforward question, but can also be a technical challenge for the engineer. The criteria for selecting the sampling location is simple, that is, to obtain a sample that represents the whole stream. However, the complicated flow patterns within a pipe, which vary with operating conditions make the selection of a constant sampling location difficult. For example, the effect of a boundary layer at a selected location may be insignificant at one operating condition, and becomes a major factor at another operating condition. There are many factors that could affect the quality of a sampling stream, only engineers with detailed process knowledge can effectively select a sampling location.

Even with a properly designed, maintained, and calibrated on-stream analyzer, the performance of the analyzer system may be compromised by a poorly designed sampling system. Often streams being

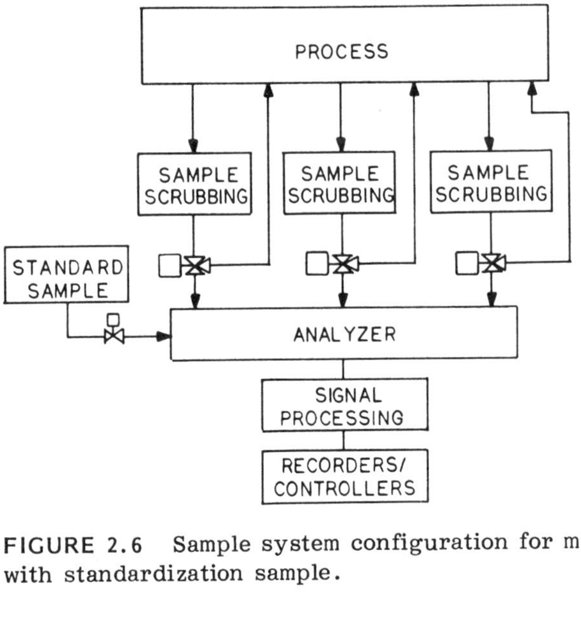

FIGURE 2.6 Sample system configuration for multistream analyzer with standardization sample.

sampled are single phase under typical operating conditions, but when drawn into a poorly insulated sample line, partially change phase, thus, causing a change in the chemical composition. The sample changes may be due to condensation of heavier components on the sample system walls causing rectification of the sample or precipitation of solids from the sample. Another critical problem, especially for samples from cracking processes, is the entrainment of minute amounts of carbon in the product stream which doesn't affect the process, but ultimately will plug a chromatograph column. These potential conditions must be recognized and the sample system designed to cope with them in order to have a functional on-line analyzer system (8).

On-line analyzers generally require frequent maintenance, not only in the sense of keeping them operating properly, but also recalibrating them against standard samples. Lacking proper and timely maintenance, an on-line analyzer may give unacceptably high errors. This is the main reason that many existing analyzers are not used for closed-loop control.

The availability of microcomputers has also helped the development of advanced on-line analyzer systems. Many new analyzer systems are being designed incorporating microcomputers to replace complicated electromechanical parts. General mechanical construction has now been simplified so greatly that analyzers may no longer be a serious maintenance problem. The most important advantage for many of these newly available analyzers, at least from the standpoint of using them for closed-loop control, is that their results can now be trusted. Self-diagnostic and automatic calibration capabilities have improved not only the reliability and accuracy, but also, the believability of these analyzers. This does not mean these analyzers do not fail. The importance is that when they do fail, they generate a fail signal to inform users. This type of smart analyzer is the most desirable device for closed-loop control.

2.3.2 Off-Line Analyzers

Off-line analyzers are generally used in quality control laboratories. Samples are collected from process streams manually and sent to the laboratories for analysis. The reasons for using off-line analyzers are many, such as lower costs and ease of maintenance as compared to on-line analyzers. In addition, some tests cannot be made on-line because of the need for destruction of the samples, such as the Mullen test for paperboard, or because of the need for special sample preparation or presentations such as the test for Mooney viscosity for synthetic rubber. Often the possibility of automatic control is ruled out when manual analysis of samples is necessary, and this conclusion is correct when only analog controllers are used for control. Computer technology has removed this limitation. Today, off-line test results can be entered into a computer automatically or manually, and the results can be used for closed-loop control. This concept will be discussed in later chapters.

2.4 DYNAMIC RESPONSE OF MEASUREMENT SYSTEMS

The potential measurement errors so far discussed have been static errors, that is, they are time-independent errors. For closed-loop control or process analysis, the response time of the measuring system must be carefully evaluated to determine the magnitude and consequences of dyanmic or time-dependent errors or the control performance using the analysis results will suffer.

To demonstrate the response of a measuring system, consider a thermocouple inside a thermowell being used to measure the temperature of a liquid stream inside a pipe. When the temperature of the liquid changes, say from 100 to 200°F, it will take a finite amount of time for the thermowell and thermocouple to reach an equilibrium with the new temperature. The time lag may be quite significant. The time-temperatures (both true temperature and measured temperature) can be expressed by a first order differential equation.

$$\tau \frac{dT}{dt} + T = T_1 \tag{2.4}$$

where:

τ = Time constant

T = Temperature as measured by the system

T_1 = Liquid temperature

t = Time

Solving Eq. (2.4):

$$T = T_1 (1 - e^{-t/\tau}) \tag{2.5}$$

Figure 2.7 shows the difference between the true temperature (assuming a step increase to 200°F at time zero) and the measured temperature, which may take many seconds to reach 200°F, depending on the type of thermowell and thermocouple used. When a measuring system is used for closed-loop control, the dynamic response should be considered when tuning the control loop, especially when the process is fast and

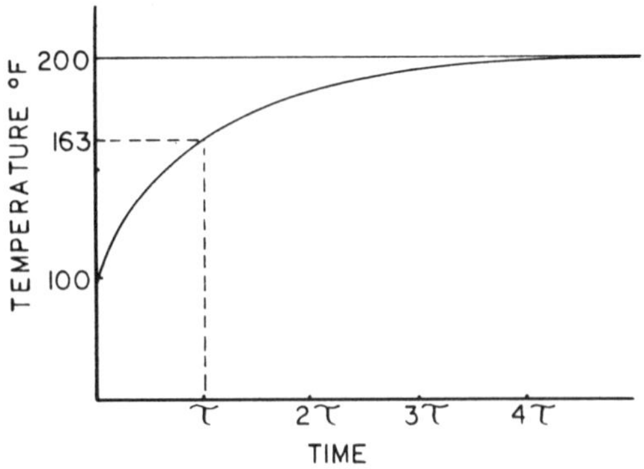

FIGURE 2.7 First order response in output signal of measuring system resulting from a step change of +100°F in the process variable being measured.

a tight control is desired. Sometimes the measurement lag can be longer than the process lag, so it may have a major impact on the performance of the control loop. Control loop tuning techniques will be discussed in a later chapter.

Dynamic response considerations of measuring a system are important for process analysis. Note that many process analysis tasks involve cause-effect evaluation between process variables. This is usually accomplished by reading the data from recorder charts. The dynamic response of one instrument could be much different from that of another, thus, when the true process response is desired, the response time of measuring systems must be considered in the data analysis.

REFERENCES

1. Lenk, John D., *Handbook of Controls and Instrumentation*, Prentice Hall, Englewood Cliffs, New Jersey, 1980.
2. Krigman, A., Flow measurement: Some recent progress, *Instrumentation Technology, 30*(4):9–13 (April, 1983).
3. Griffiths, C., A New Look at Flow Measurement Transducers, *Pipeline Industry*, March, 1982, pp. 77–84.
4. McCabe, W. L. and Smith, J. C., *Unit Operations of Chemical Engineering*, McGraw-Hill, New York, 1967.
5. Demorest, Jr., William, J. There's more to transmitter accuracy than the spec., *Instruments & Control Systems, 30*(5):61–63 (May, 1983).

6. Laduzinsky, A. J., Sensors Employ Micros for Added Functionality, *Control Engineering, 30*(6):70–72 (June, 1983).

7. Stroud, L. J., New transmitter speeds process operations at Southwest Refining, *Control Engineering, 31*(8):200–206 (August, 1984).

8. Agarwal, A. K., Zielinski, R. E., and Seabaugh, R. W., "On-line analysis and sampling of process and product variables for the Pricetown I Underground Coal Gasification Field Test," Proceedings of the 1982 Symposium on Instrumentation and Control for Fossil Energy Process. Research report by the Monsanto Research Corporation for the U.S. Department of Energy under Contract No. DE-AC04-76-DP00053. Argonne National Laboratory, 1982.

3

Conventional Control Systems

3.1 THE PROPORTIONAL-INTEGRAL-DERIVATIVE CONTROL ALGORITHM

The major function for an analog controller is to hold a process variable at a manually entered target value or setpoint. As shown in Fig. 3.1, the controller receives a measured signal of the controlled process variable, compares it with the setpoint value, which generates an error value, and outputs a signal, which is a function of the error value, to an actuator. Ideally, the output correction to the actuator will result in an exact correction in the process variable, forcing it to once again equal the setpoint. Those factors which disturb the controlled process variables are disturbances. Without disturbances, there would be no need for measurement nor control of a process.

To clarify this discussion of conventional control systems, the following definitions will provide a better understanding. The variable being measured and held at the setpoint is defined as the controlled variable, x, while the controller output that manipulates the process is defined as the manipulated variable, u. The controller which computes an output based on the error signal is defined as a feedback controller (1).

Note that the feedback error is computed by subtracting the measured value of the controlled variable, x, from the setpoint value. The feedback, in this case, is always negative with respect to the setpoint value. This is the nature of negative feedback that of forcing a process to a stable condition.

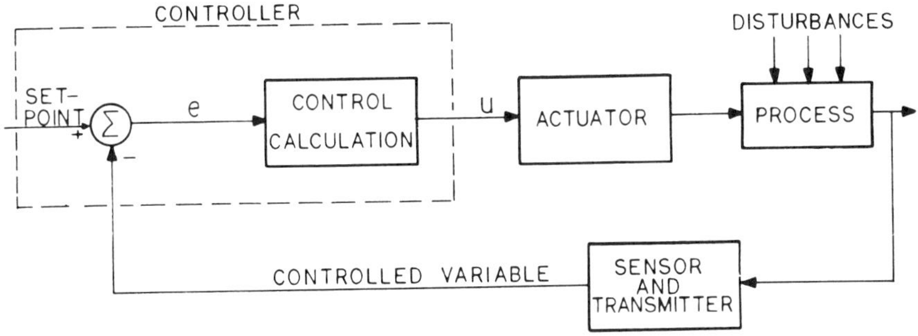

FIGURE 3.1 The controller reads the signal from the transmitter and sends an output signal to the actuator. The actuator output and disturbances affect changes in the process or the controlled variable, which is measured by a sensor, and its value is transmitted to the controller perpetuating further control actions.

Like the process measuring devices discussed in the previous chapters, early controllers used for process control were pneumatic controllers. These controllers were analog devices, that is, they received a continuous process measurement signal and generated a continuous output signal to the actuator.

Figure 3.2 shows the design concept of a proportional only controller. In this case, the error is converted to a corresponding mechanical motion. This motion is then translated into a pneumatic signal, depending on the gain adjustment and the manual reset positions. The mathematical expression for this device is as follows:

$$u = k_p \times e + u_0 \tag{3.1}$$

where

u = Controller output

k_p = Controller gain

e = Setpoint value − feedback measurement value

u_0 = Controller output bias, defined as manual reset

Every variable is assumed to have a range of 0 to 100 percent. The adjustable proportional constant, k_p, can be set at between zero to approximately ten, depending on the type of application.

Equation 3.1 shows that the output of a proportional controller is the sum of $k_p \times e$ and u_0. Under special conditions, for this controller,

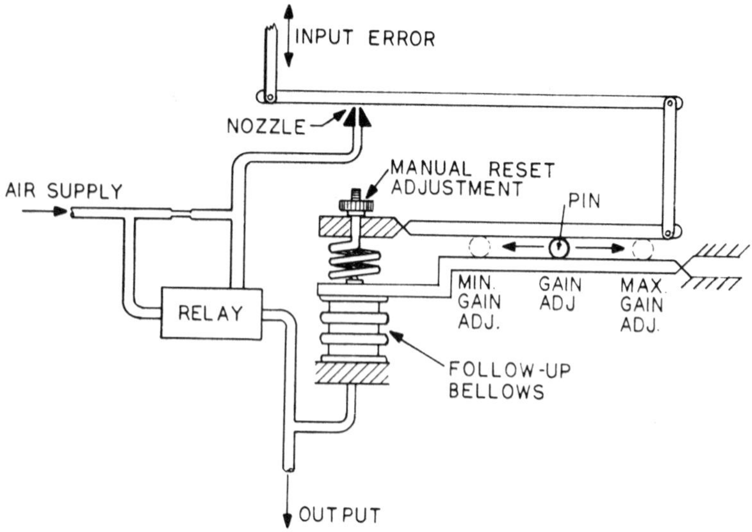

FIGURE 3.2 A pressure imbalance (error) is generated from the setpoint and feedback. When the baffle is pushed against the nozzle by the pressure imbalance, the direct acting relay and the bellows increase the output until the input error is changed to the other direction (through a change in the process). The gain adjustment determines to what extent the error signal is amplified. The manual reset adjustment determines the magnitude of the force that controls the output required to hold the controlled variable at the setpoint, that is, it cancels the pressure balance.

the controlled variable (output) will oscillate around the setpoint, and the amplitude of oscillation is determined from the initial condition. This special condition is when the loop gain function for the process and the controller is set close to negative unity.

The term, $k_p \times e$ is computed from the value of e and the setting of k_p. Some manufacturers substitute the term proportional band (PB), which is defined as the percent in error changes resulting from a 100 percent controller output change, to replace the constant k_p mathematically,

$$PB = \frac{1}{k_p} \tag{3.2}$$

Figure 3.3 shows a graphic procedure to compute a controller output. Assume the controller gain is set at 0.5, then at a given time when the setpoint is 50 percent and the feedback is 60 percent, the controller output will be (0.5 × (50−60) + 50) or 45 percent.

Since u is a continuous output that changes the process, the controlled variable, x, changes with u. This forms the closed-loop relationship between u and x. Figure 3.4(A) shows a simple time delay process where the controlled variable, x, follows the manipulated variable, u, with the same magnitude and pattern, only at a later time (τ seconds) period. Figure 3.4(B) shows the relationship between u and x during a period after a step change in septoint is made. When the proportional constant, k_p, is set at less than 1.0, say at 0.5, the controller output, u, will oscillate around u_0 with the decaying amplitude, and the controlled variable, x, will be stabilized around the setpoint. Again, this is under ideal conditions. In reality, the load of a process change and the controlled variable may not be linear with relation to the manipulated variable. In a nonlinear process, equal changes in the manipulated variable at different absolute conditions do not have equal effects on the process variable. These nonideal conditions cause a proportional only controller to be inadequate for controlling some processes because it will result in process response offset. The offset of a process response caused by a proportional controller

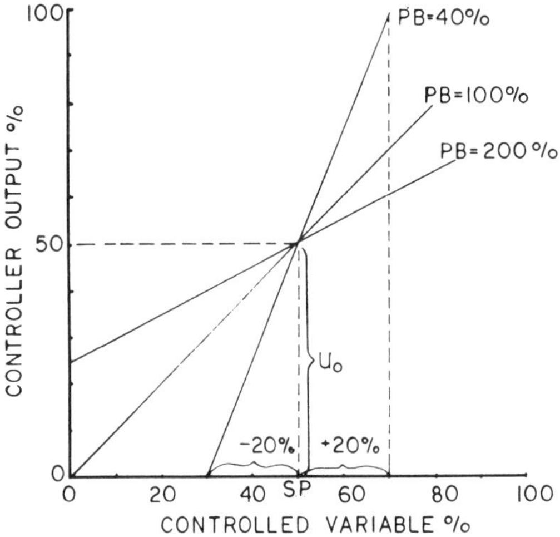

FIGURE 3.3 Graphics for computing controller output at different proportional bands and a constant 50 percent manual bias. When PB = 40%, a 40 percent error changes the controller output by 100 percent; when PB = 200%, a 40 percent error changes controller output by 20 percent.

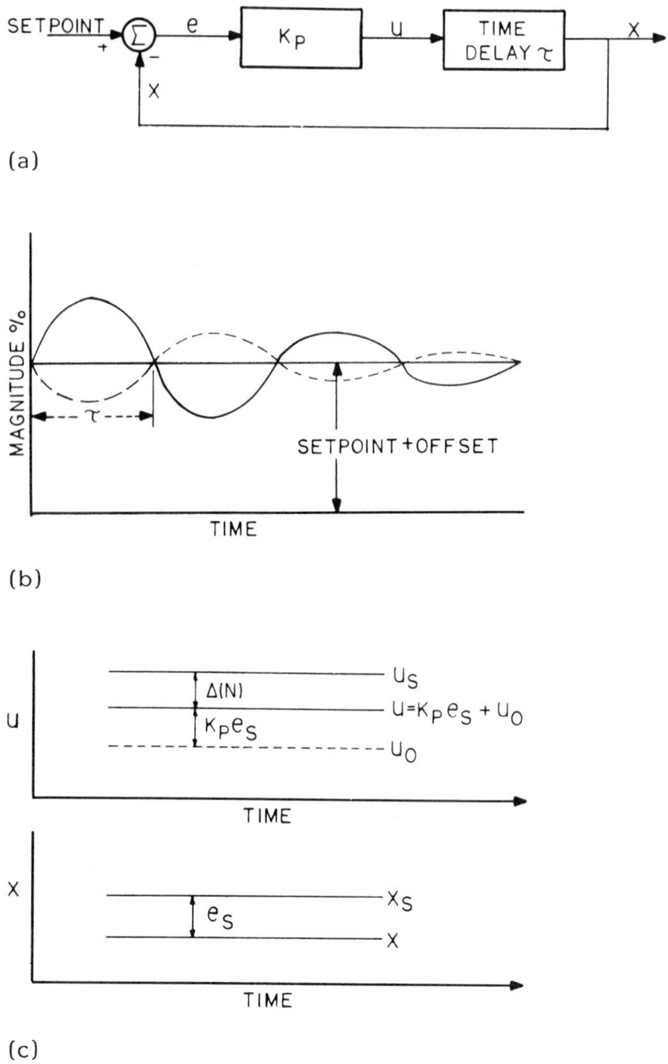

FIGURE 3.4 Controlling a simple time delay process by a proportional controller. At 200 percent proportional band, the process variable experiences an oscillation with reducing amplitude. (a) Control loop; (b) relationship between x and u when $k_p = 0.5$; (c) offset relationship between controlled process having time delay.

can be computed from the following equation, and can be observed from Fig. 3.4(c):

$$e_s = \text{offset} = \frac{u_s - u_0 - \Delta(N)}{k_p} \tag{3.3}$$

where u_s is the steady-state controller output value which will drive the controlled variable to the setpoint. The value of u_0 can differ from that of u_s due to two factors: first, u_0 is usually set at one value while the setpoint varies with operating conditions, and second, the process itself may change requiring a different controller output value to reach the same setpoint. One example of the second factor is where a control valve at the 50 percent open position may produce a 50 percent flow rate at one time, and may produce a 60 percent of flow rate at another time because of a higher header pressure. The combined effect of these process changes is defined as $\Delta(N)$.

Regardless of the causes, the offset of a proportional controller changes with the settings, u_0 and k_p. At a given condition, a higher controller gain, k_p, results in a lower process offset, however, greater controller gain also causes greater oscillation. Thus, proper tuning of a controller is a compromise between two undesirable conditions. To eliminate this need for compromise, a second control function, reset, is added to eliminate the undesirable offset condition.

To correct the offset caused by a proportional controller, an automatic reset element is added. For a pneumatic controller, as shown in Fig. 3.2., this is accomplished by replacing the manually adjustable spring with a bellows, and the air flow to the bellows is manually adjustable to speed up or slow down this reset. An identical function is implemented for electronic controllers using amplifiers. The mathematical expression for a proportional plus reset controller is:

$$u = k_p \times (e + k_I \int e\, dt) \tag{3.4}$$

where k_I is the reset constant; a higher value of k_I gives faster reset actions, and its unit is repeat per minute.

The term repeat per minute, can be explained by the example shown in Fig. 3.5. At time zero, a constant error of 10 percent is applied to the controller. According to Eq. (3.4), assume k_p is set at 2, and k_I is set at 1. At constant error, $\int e\, dt$ becomes $e \int dt$, or $e \times t$. Thus,

$$u = k_p e + k_p k_I \times e \times t$$

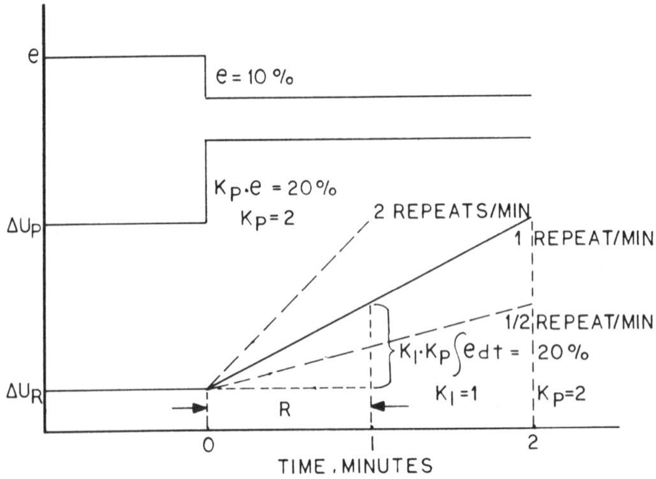

FIGURE 3.5 The response of a proportional and integral controller to a constant error. (This is just a simulated condition for demonstrating the controller behavior at a defined condition. In reality, the error should not be constant because of closed-loop control.) The proportional output is inversely proportional to the error, while the integral output increases with time.

or

$$u = 2 \times 10 + 2 \times 1 \times 10 \times t$$

where t = time measured in minutes.

After one minute, the change in controller output contributed by the reset action is 20%, or just a repeat of that provided by the proportional mode. That is, when k_I equals 1, it implies one repeat per minute.

As shown in Fig. 3.5, when $k_I = 2$, the output change contributed by the reset action after one minute is twice as much as the change contributed by the proportional mode, thus, two repeats per minute. Because this reset action is based on the integration of errors, it is also called integral mode. Thus, the proportional plus reset controller is also called proportional plus integral, or the PI controller.

Most of the analog controllers in use are PI, proportional plus integral, type controllers. Some manufacturers use a constant, (1/R) or $(1/T_I)$, to replace k_I, and R is defined as reset time in minutes per repeat. In this case, greater values of R or T_I give slower integral (or reset) actions. The values of the tuning constants k_p and k_I can be determined by the process model as discussed elsewhere.

As more and more processes demand automatic control, PI controllers have been considered too slow for controlling a process which experiences sudden upsets, i.e., large disturbances, or a batch process with large step changes in setpoints. To speed up the controller responses to such disturbances, a rate or derivative mode is added. The mathematical expression for the derivative mode is:

$$\Delta u_D = k_D \frac{de}{dt} \tag{3.5}$$

where k_D = derivative constant in minutes.

Figure 3.6 shows the derivative controller output, Δu_D, when de/dt is a constant. When the tuning constant k_D is set at 1.0, the controller output leads the error by one minute, so the unit for the constant, k_D, is minutes.

The three mode, or proportional-integral-derivative (PID) controller is the combination of Eqs. (3.4) and (3.5).

$$u = k_p\left(e + k_I \int e\, dt + k_D \frac{de}{dt}\right) \tag{3.6}$$

The above equation is for the PID control algorithm used for the design of both pneumatic and electronic analog controllers (2). In fact, the most widely used control algorithm for digital controllers is just a digital form of the above equation, as is to be discussed in later chapters.

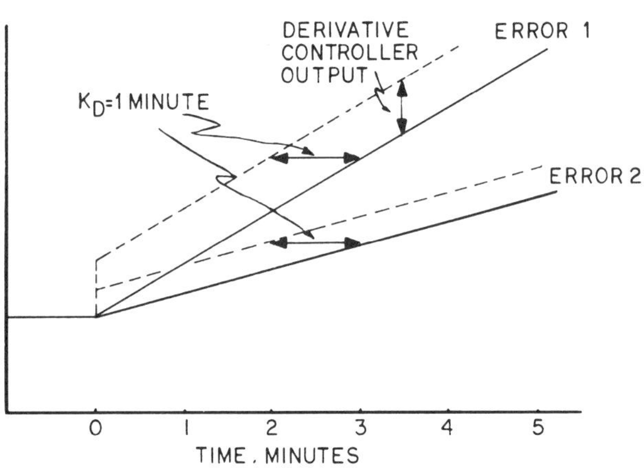

FIGURE 3.6 The response of a derivative controller to errors at constant slope. (This is also a simulated condition for demonstrating the controller behavior.) When the tuning constant, k_D is set at one minute, the derivative controller output leads the error by one minute.

The tuning of a controller, i.e., to find the optimum setting for the constants, k_p, k_I, and k_D for a given process, is not as simple a task as some people might expect, and this technique will be discussed in later chapters. Generally speaking, a controller should be tuned according to the type of process (the process model) and the desired control performance. In addition, consideration must be given to the disturbances which are expected to influence the process and operating condition changes required by the product mix.

3.2 COMMERCIALLY AVAILABLE ANALOG CONTROLLERS

3.2.1 Single-Loop Controller

Like transmitters, there are two basic types of analog controllers: pneumatic and electronic. The standard input/output signal range for a pneumatic device is 3 to 15 psi of air pressure, while that for an electronic device is 4 to 20 ma DC current. Many models of each basic type of controller have been produced over the many years. Although most of them are based on the PI or PID algorithms as discussed above, each model offers somewhat different capabilities. Users need to be aware of the fact that a controller is more than a control algorithm. The ease of operation and communicating with other devices, such as another controller or a computer, is also very important.

Figure 3.7 shows a simplified block diagram of a commercially available PI or PID controller. It is not based on any particular design, and is only used to demonstrate the factors that users should be concerned about when selecting a specific controller for a control application.

As shown in Fig. 3.7, an analog controller has a Manual/Auto switch, M/A, two display devices, two manual setting devices, and one control algorithm computing device. In addition, the three tuning constants are adjustable by turning screws or other hardware connected to the computing device. All three displays, process variable, setpoint, and controller output, are usually shown as 0 to 100 percent scale. Some controllers allow users to change displays to engineering units such as 0 to 300 GPM.

The M/A switch is used to open/close the control loop, that is, the control algorithm output signal is disconnected from the control actuator device. Manual setting of controller output is necessary when a process is first started, when the sensor or input signal is bad, or when a fast change of the process is required. The switching from manual to automatic, or from automatic to manual, can be a simple operation or a more complicated operation, depending on whether the controller is designed for bumpless transfer.

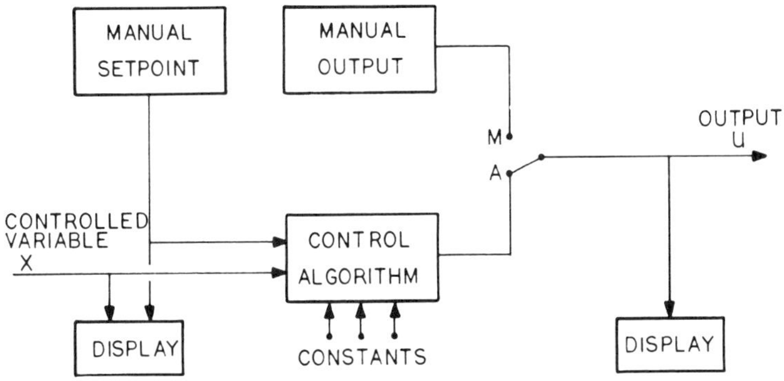

FIGURE 3.7 The source of the controller output, u, can be from either the manual output or the control algorithm computation blocks, depending on the manual/auto (M/A) switch position. In order to have a bumpless transfer from M to A, the signal level at both A and M positions must be equal. An automatic bumpless transfer circuit can be implemented by using the integral amplifier function in the control algorithm as the manual output holding device. In this case, the setpoint still must be manually adjusted to the value of the controlled variable or else the integral mode controller will start moving the output as soon as the switch is turned.

The process could be bumped when the switch is turned from "M" to "A" or "A" to "M". This happens when the output from the control algorithm calculation does not equal that of the manual output setting. Before the bumpless transfer design was implemented, it was the operator's job to balance these two output signals manually before any switching could be made.

Some controllers offer bumpless transfer from "M" to "A". This is accomplished by setting the integral output equal to the manual output when the switch is turned to "M". Under this condition, the operator should still set the setpoint equal to the controlled variable before he turns the switches to "A", or the process could be bumped as soon as the switch is turned because of the error between the setpoint and the value of the controlled variable. Manual balancing of the output is needed for most conventional analog controllers before switching from "A" to "M".

Another standard feature for new controllers, which was not available for many conventional controllers, is the antiwindup function. Windup is the condition where the integrator used for integral control becomes oversaturated and to unwind this integrator requires a prolonged time. This is undesirable because the controller output cannot be changed for a period of time. Antiwindup is a design to prevent oversaturation of the integrator so the output can be reversed whenever the control algorithm calls for it.

3.2.2 Controller with Remote Setpoint

A special add-on feature for some controllers is the ability to accept a setpoint input from another device, such as a controller or a computer. This is accomplished by adding a remote/local, R/L, switch and remote setpoint generator to the controller as shown in Fig. 3.8. Three types of designs have been utilized for this remote setpoint generator.

The most popular device is one that receives a remote setpoint as a continuous analog signal at the same range of the controller output. The remote setpoint generator just transmits the remote signal to the controller. Both pneumatic and electronic controllers offer this type of device.

Another type of device used for accepting remote setpoints is designed to receive an electronic signal for a short period of time and hold the magnitude of that signal while the remote signal is disconnected. The setpoint generator in this case is a sample-and-hold device. This is mainly designed to permit interfacing a number of controllers to a supervisory computer by using just one digital-to-analog converter. As the cost of digital-to-analog converters becomes lower, the application of the sample-and-hold becomes unnecessary, but many existing supervisory control systems use these types of analog devices.

An even older device for generating a setpoint is a stepping motor driven analog signal generator. This type of device has been used for

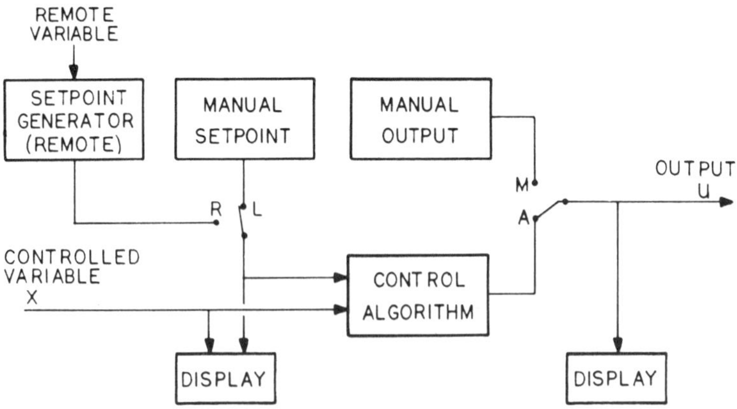

FIGURE 3.8 Identical to Fig. 3.7 except a Remote/Local (R/L) switch is added for selecting the source of the setpoint.

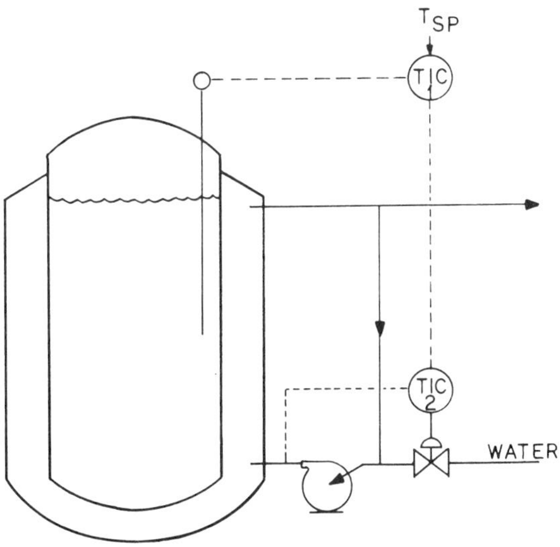

FIGURE 3.9 The jacket water temperature controller (TIC-2) is called the inner (or slave) loop while the reactor temperature controller (TIC-1) is called the outer (or master) loop. TIC-2 will reduce the effect of the disturbances, such as the water temperature variations, before they impact the reactor temperature.

interfacing supervisory computers with both pneumatic and electronic controllers. The design concept is to allow an analog controller to accept its setpoint in the form of electronic pulses. The full range of the setpoint value is usually divided into 1,000 steps. A computer usually has the capability of driving the setpoints of a number of controllers using a single pulse output controller.

3.2.3 Cascade Controllers

Cascade controllers are used to control a process where an intermediate variable can both be measured and controlled. The purpose for the control of an intermediate variable is to reduce the influence of disturbances on a process before they have any effect on the controlled variable. Figure 3.9 shows a typical example for the application of cascade controllers used to control the temperature of a reactor. The temperature controller, TIC-2, controls the jacket water temperature by manipulating the water control valve, and the temperature controller, TIC-1, controls the reactor temperature by manipulating the setpoint of TIC-2. This arrangement will provide better control performance than that experienced from a single controller configuration, connecting the

TIC-1 output directly to the control valve and eliminating TIC-2. Cascade control provides superior performance because the disturbances to the jacket temperature are reduced before they can have any effect on the reactor temperature.

To close a cascade loop without causing a bump requires several manual operations unless process variable tracking and automatic initialization capabilities are provided for the controllers. To demonstrate these manual operations, Fig. 3.10 shows the linkages between the two controllers.

Assuming the process is at manual control (all switches are on "M" or "L"), and the operator wants to close the cascade loop, he must perform the following operations:

1. Set the TIC-2's manual setpoint equal to the controlled variable, T2, and then turn the TIC-2 A/M switch to "A".
2. Set the TIC-1 manual output equal to the TIC-2 setpoint, and turn the TIC-2 L/R switch to "R".
3. Set the TIC-1 setpoint equal to the controlled variable T1, and turn the TIC-1 M/A switch to "A".

The above operation can be somewhat simplified when the R/L switch of the TIC-2 is kept at the R position all of the time. Operators can

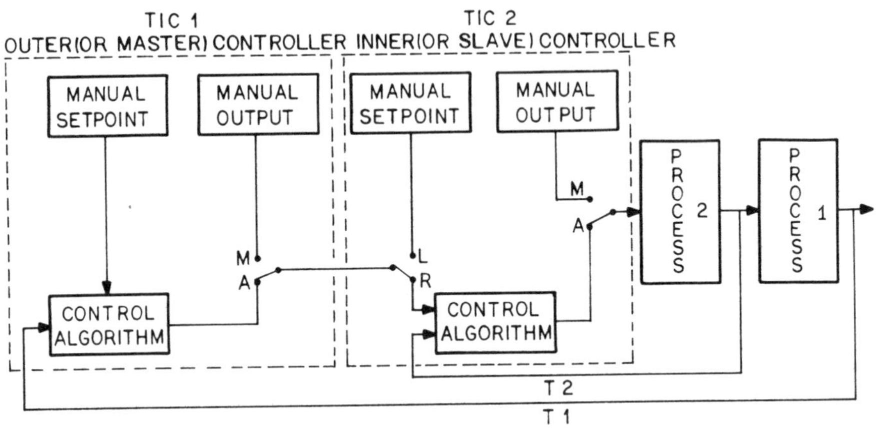

FIGURE 3.10 Process 2 includes the water control valve, the pump, and the pipe between them, i.e., the jacket water temperature responded to the control valve position. Process 1 includes the reactor itself, i.e., reactor temperature responded to jacket water temperature. The R/L switch of TIC-2 can remain at the R position because the setpoint of TIC-2 can be changed manually by using the manual output signal from TIC-1.

initialize the inner loop by using the manual output knob of the outer loop. Even so, the operation still requires the operator's time and attention, and the operators may prefer to maintain manual control for a prolonged period. This makes the controlled variable tracking and automatic initialization functions very desirable.

3.2.4 Other Analog Control Devices

Over the years, complicated control loops have been developed using analog controllers and computing devices. Most of these developments have been based on experiments linking controllers and computing devices together to perform sophisticated control functions such as feedforward control. These control functions will be discussed in later chapters.

Basic computing devices such as the summer and multiplier, and logic devices, such as the high selector and low selector, have been provided for the design of control functions.

Using analog devices for implementing sophisticated control strategies requires implementation skills which are unfamiliar to the typical process engineer. This is one of the reasons why process engineers have not typically been involved in the design of control systems. The implementation of sophisticated control functions using analog devices can be demonstrated by the example shown in Fig. 3.11, which is a simplified process control flow diagram for the control of a boiler drum level. This is a so-called feedforward-feedback control system, which is also called a three-element boiler drum level control function, indicating that the drum level is controlled by measuring three variables (the level, feed water flow rate, and steam flow rate). The design concept is that when the steam flow is changed drastically, due to changes in process demand, the level controller alone may not act fast enough to correct this change, so the steam flow is used as a feedforward signal to change the boiler feed water flow directly.

Figure 3.11 shows one of the many ways to implement three-element drum level control. In this case, the linearized steam flow signal is combined with the level controller output by a "summer," and then this value is used as the setpoint of the boiler feed water flow controller. Note that the summer does not work as if it is adding two numbers together, it only works when both of the inputs are properly scaled. Both inputs and the output of the summer must be within the range of 0 to 100 percent (3 to 15 psi or 4 to 20 ma). Tuning constants must be applied to the two inputs according to the following operations:

$$k_1 \times F + k_2 \times L = F_{SP} \tag{3.7}$$

$$k_1 + k_2 = 1 \tag{3.8}$$

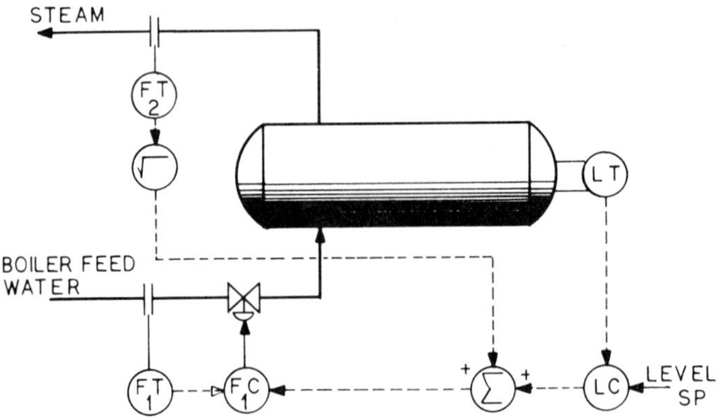

FIGURE 3.11 One way to implement the three-element boiler control function. The steam flow is used as a feedforward signal for the level control. The purpose is to increase (or decrease) the boiler feed water flow during a sharp increase (or decrease) in steam demand. This function reduces the disturbance caused by steam flow changes on drum level.

where

F = Linearized steam flow signal, %

L = Level controller output, %

k_1, k_2 = Adjustable constant between 0 to 1.0

The constants, k_1 and k_2, are determined by the control system engineer based on the actual conditions of the boiler. It is more or less an art because no real scientific rules can be applied. Note that when k_2 equals 1.0, the control system becomes a feedback-only cascade controller.

The addition of a feedforward element to a control loop does not always produce a positive result. Using the control system shown in Fig. 3.11 as an example, it may have the following two potential problems as compared to a feedback only control system.

First, closing the loop could be difficult for operators, because two levels of manual adjustments are needed in order to have a bumpless transfer. Second, the control loop becomes less reliable because three more hardware devices are added to the basic level-flow cascade loop. These are the flow transmitter, the square root extractor, and the summer. Because of the concerns discussed above, the maintenance of sophisticated analog control systems requires adequate control system expertise (3).

3.3 INTERFACE ANALOG CONTROL SYSTEMS WITH SUPERVISORY COMPUTERS

Many existing control systems in process plants are based on analog devices. To evaluate the feasibility of adding a supervisory control computer to improve control performance, the technique and cost for interfacing the existing analog control systems and the computer is a major concern. Note that the term *existing analog control systems* is used because all new control systems which may need a supervisory computer should incorporate digital control devices which are to be discussed later.

Both electronic and pneumatic control systems can be interfaced with computers (4). The most reliable system at an acceptable cost should be used. This may mean to add devices to the existing systems, or to replace the existing system with new devices.

Figure 3.12 shows a classical way to interface a pneumatic control loop with a supervisory computer. This technique was widely applied during the early years of computer application when only a limited number of control loops were supervised by one computer. Note that two pressure-to-current transducers are used for each loop; one for the process feedback and the second for the setpoint feedback, which must be read by the computer to identify whether the stepping motor

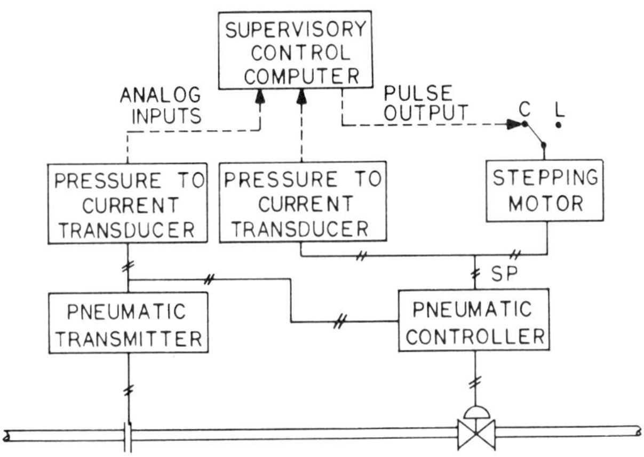

FIGURE 3.12 Two pressure-to-current transducers are needed to interface a pneumatic control loop to a computer, provide the controller has a stepping motor driven setpoint. The setpoint should be read back to the computer to identify the position of the setpoint and to identify problems, such as controller offset or the failure of the stepping motor. A relay is usually needed to provide a contact to indicate the C/L switch position.

is functioning properly or whether the operator has changed the setpoint during the computer control mode. This technique is rarely used now because few computers support pulse output controller as a standard device.

Figure 3.13 shows another method to interface a pneumatic control loop to a supervisory computer. A current-to-pressure transducer is used for transmitting the computer output to the controller, and a pressure-to-current transducer is used for tracking the setpoint, so a bumpless transfer from local to computer can be made. This is accomplished by reading the manual setpoint when the Computer/Local (C/L) switch is turned to L, and keeps the computer output equal to the manual setpoint.

Figure 3.14 shows how an electronic control loop can be interfaced with a computer. The only requirement in this case is that the controller must be able to accept remote setpoints. The setpoint feedback is not needed in most cases, because electronic controllers have a built-in tracking circuit.

In addition to the analog signals that must be interfaced with the computer, the two switches of each analog controller, i.e., the statuses of the computer local (C/L) and the auto/manual (A/M) switches, must also be read by the computer as digital inputs. This information is necessary to keep the computer on the right track. For example, the

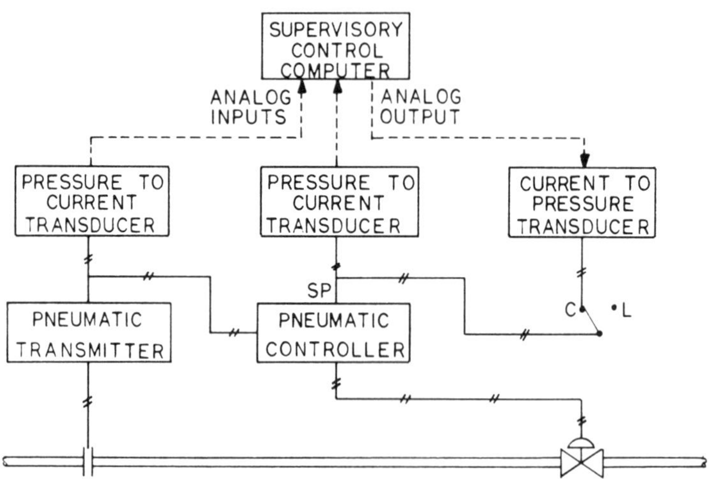

FIGURE 3.13 A pneumatic control loop with remote setpoint capability can be interfaced to a computer using two pressure-to-current transducers and one current-to-pressure transducer, plus a relay for the C/L switch.

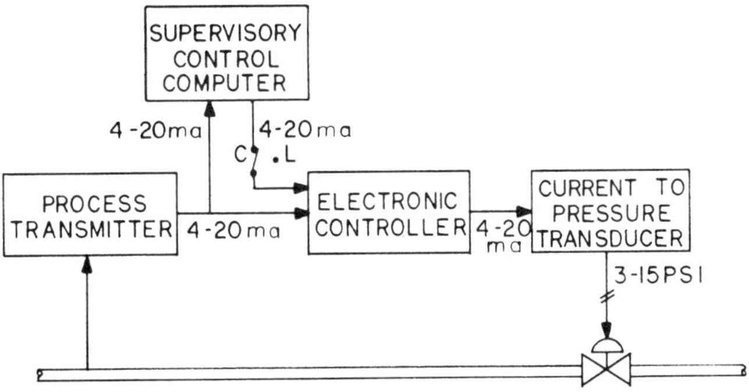

FIGURE 3.14 An electronic control loop with remote stepoint capability can be interfaced to a computer without transducers.

computer control function should be initialized when the C/L switch is first turned to C.

Based on the above discussions, interfacing existing analog control systems to a supervisory computer can be costly. This is particularly so when the total number of loops are high. When this is the case, consideration should be given to replace the existing analog control system with a new distributed control system (this will be discussed in Chapter 4), which can be interfaced with supervisory computers by data links.

REFERENCES

1. Perry, Robert H., *Chemical Engineer's Handbook, 6th Edition*, McGraw-Hill, New York, 1984.

2. Considine, D. M. and Ross, S. D., *Handbook of Applied Instrumentation*, McGraw-Hill, New York, 1964.

3. Shinskey, F. G., *Process Control Systems*, McGraw-Hill, New York, 1967.

4. Merritt, R., "Interfacing digital electronics with pneumatics," *Instruments & Control Systems*, *56*(4):47–52 (April, 1983).

4

Digital Control Systems and Programmable Logic Controllers

4.1 INTRODUCTION TO DISTRIBUTED CONTROL

The year 1976 witnessed the technological giant step forward for instrumentation and control when Honeywell announced and introduced: the *microprocessor based digital controller*, which knelled the demise of the conventional analog controller; *the concept of the data highway*, making obsolete current instrumentation design and installation practices; *and the versatility of shared display operator consoles*, replacing the expensive rigid control graphic panels, indicators/recorders, and annunciators (1). The introduction of the distributed digital control concept has allowed an entirely new approach to control system design for the process industries.

Before distributed control, most computers were used as special add-on equipment to an existing control system, such as data logger, management information system, or supervisory control. There were a few computers installed as centralized Direct Digital Control (DDC) systems, that is, using one computer system (containing two to three central processing units) to replace all control room instruments including controllers, recorders, annunciators, etc. However, these early DDC systems did not have much success because of the risk of losing all controllers at the same time, and the high cost of installing computers and backup systems. Distributed control systems overcome all these problems. The distributed control concept involved three basic ideas:

1. Replacement of the local analog instrumentation with sophisticated, easily configurable, microcomputer based controllers which perform

the local control functions. These are distributed throughout the process and are located in close proximity to the equipment being controlled. Each basic controller is a stand-alone, preprogrammed computer containing a library of some 50 control strategies. The process engineer simply configures the desired control strategy for a particular area from this library. If at any time a new strategy is desired, the basic controller can be easily *reconfigured*. The controllers also permit logical decision; no programming is involved (1).

2. A redundant, high-speed communications highway which transmits process data from the local controllers to the operators' consoles in the centralized control room using a reliable digital format. Operator-initiated control actions or inquiries are also communicated to the microcomputers via the data highway (1).

3. Operator interface consoles containing multiple colored CRT displays which are driven by microcomputers. This permits the presentation of a wide variety of informative, sophisticated displays to the operator. These range from data being presented in tabular form to trend or as graphic displays representing live process data. Special keyboards with functional keys are provided to operators for easy access to the data base (2).

Distributed control systems have enjoyed unprecedented success in the instrumentation and control business. As of 1984, more than 30 vendors offered distributed control systems. A great majority, if not all, of new large plants are being built using distributed control systems.

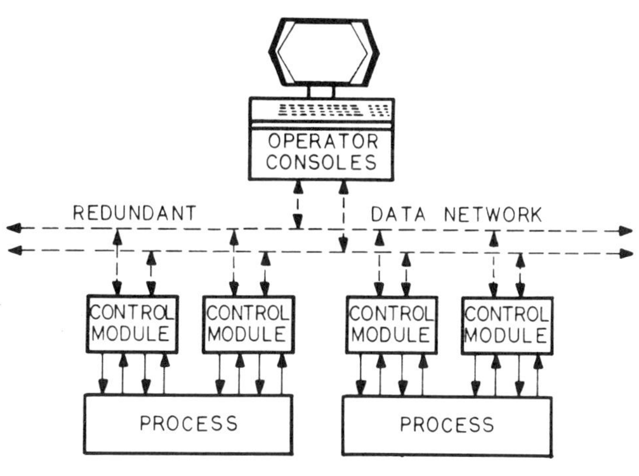

FIGURE 4.1 Distributed digital control system consists of microprocessor-based digital controllers distributed throughout the process(es) connected to a central CRT based operator's console via a digital communications network.

Each vendor designs his system differently from the others. The most significant difference in design concept is the number of conventional loops to be replaced by one microcomputer. This ranges from one loop per computer to somewhat over 50 loops per computer. There are good reasons to support both cases. A user needs to choose a system from the needs of his application.

A typical configuration of a distributed control system is shown in Fig. 4.1.

4.2 DISTRIBUTED CONTROL ARCHITECTURE

The architecture of the distributed control system can be varied or customized to meet the specific requirement for control of a particular process area. In its simplest form, the control system can be designed around a single control module which can be installed locally and perform a dedicated function without transmitting any information to a central control room. The operator interface would be the controller faceplate display, which can range in complexity from a simple gauge showing the value of the process variable, and several push buttons to adjust the controller setpoint and change the controller status, to a considerably more sophisticated faceplate display which includes Light Emitting Diode (LED) bar graphs and extensive use of alphanumeric displays to provide readouts in engineering units. However, the complexity of the control algorithm is not compromised, even in this simplest configuration, in that typically up to four analog inputs, two analog outputs and several digital inputs and outputs, can be incorporated into a complex control strategy utilizing the full complement of controller functions modules (3,4).

In the case where a single microprocessor-based controller supports multiple control loops (8–32), operator communications can be accomplished via a data entry panel consisting of keyboards and LED-type displays which will provide displays for the controller status on a loop request basis, or via a locally installed Cathode Ray Tube (CRT) display with keyboard. Group displays would be possible when the latter alternatives are used.

Field instrumentation signals are terminated in the same racks containing the controller module. Therefore, there is considerable economic incentive to locate the controller module rack close to the process in order to minimize the cost of field wiring. This type of installation is shown in Fig. 4.2.

The next level of sophistication would be the use of a distributed control system to replace a conventional central control room. In this case several controller modules are required and possibly a large number of noncontrol related analog and digital input modules are desired.

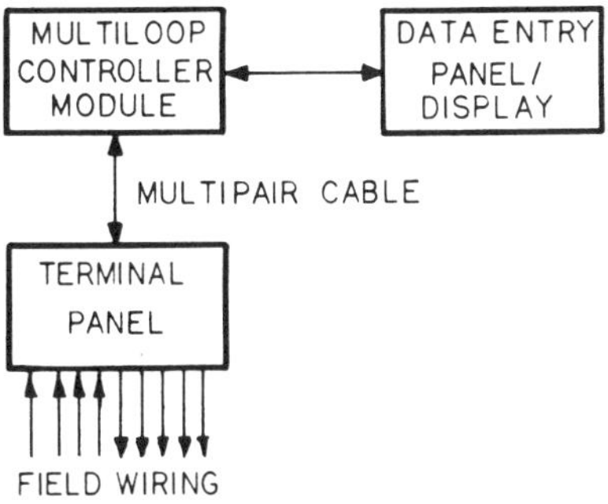

FIGURE 4.2 Installation of local multiloop digital controller via a terminal panel which provides termination points for field signals. Digital controllers can be equipped with data entry/display panels for operator communications.

All control and input modules would be connected to a digital communications network (called data highway, data bus, bus data highway, data freeway, etc.), and would share an operator's console, which would also reside on the highway. An example of this configuration is shown in Fig. 4.3 (5).

The function of the process interface unit would be solely to convert analog input signals to digital equivalents and to make these available for polling by the network. Some input signal checking might also be performed by these devices.

The process information would be displayed on the operator's console, which consists of CRT, keyboard, and maybe some special displays and switches, in response to a request initiated by the operator. Control module status, setpoint, and tuning constant changes would be initiated by the operator and communicated to the respective controller over the network. This is shown in Figure 4.3.

The capability for communications across each vendor's data communication network (bus) varies greatly. In all cases, the operator console polls the communications network on a frequent basis to determine the status of each of the network modules and to obtain requested data. In some cases, individual modules may initiate data transfers to other modules on the same network. This is particularly beneficial when it is desired to build complex control strategies incorporating cascade and feed-forward control, and control values need to be passed between controllers.

It should be mentioned here that because of the extreme importance of the reliability of the data communications network, in that it is the sole connection between the operator and the process, systems are almost always configured with totally redundant communications channels.

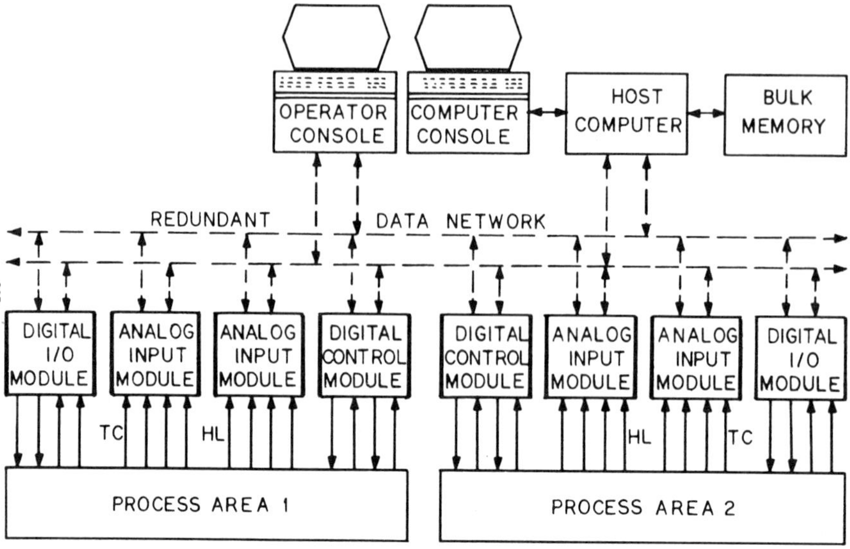

FIGURE 4.3 Centralized control with distributed control system is accomplished via locating various functional modules in close proximity to process being controlled, and communication between these models and the CRT based consoles is accomplished over redundant data networks. The host computer system also resides in the same network. The modules include those for digital input/output signals, low level thermal couple input signals, high level process input signals, and digital controllers.

The next level of design complexity for the distributed control system can follow one of several alternatives and does not just simply add more of everyting. One of the most frequent alternatives is to add a host computer to enhance control system capabilities. This level of control now permits the implementation of complex supervisory control strategies, integrated or coordinated control, and complex start-up/shutdown sequences. Another alternative is to add advanced control modules to the system configuration. These are modules that support comprehensive mathematical calculations, batch or sequential control, programmable logic control, etc.

As the distributed control system configuration grows in complexity, so does the complexity of the data communication network increase. The typical network will evolve from a single dual data bus to multiple dual highways or freeways in various configurations. The configurations most frequently seen are "star", "T", "loop", and parallel hierarchy.

Figures 4.4 and 4.5 show how a complex distributed control system might be configured (3,6,7).

4.2.1 Process Connected Modules

In traditional process management systems, a maze of dedicated wires and junction boxes connect field sensors, transmitters, and actuators to centrally located control and data acquisition equipment. This traditional method of centralized control leaves the plant vulnerable to accidents that might sever communications between controllers and the process. Also, the number of wires required for such systems adds substantially to installation costs and space requirements.

The distributed control system simplifies wiring, reduces installation costs, and increases process security by distributing control and data acquisition equipment close to the process. All equipment is microprocessor-based and functions independently; hence, no device depends on another for performance. Though fully distributed, process

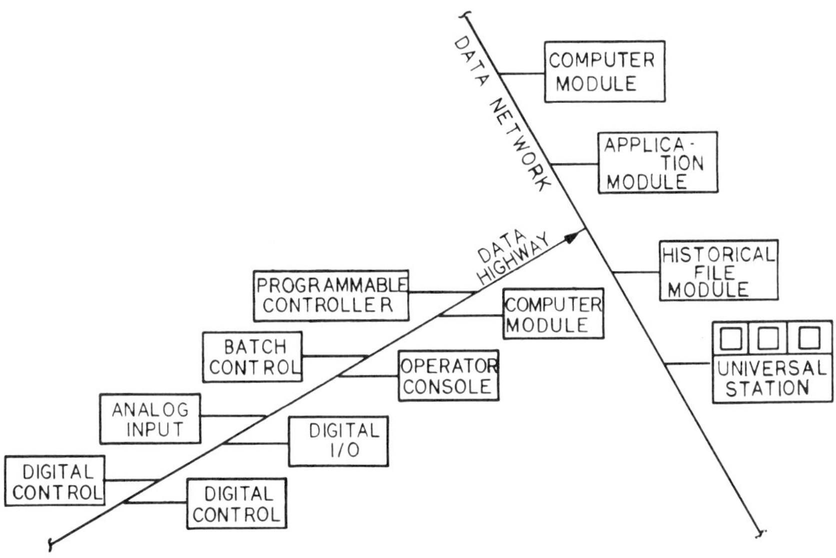

FIGURE 4.4 One alternative approach for configuring a distributed control system is where multiple local data highways are connected to the high-level data networks in a T configuration. The high-level data networks provide high-speed interprocess communications capability and access to share computing resources.

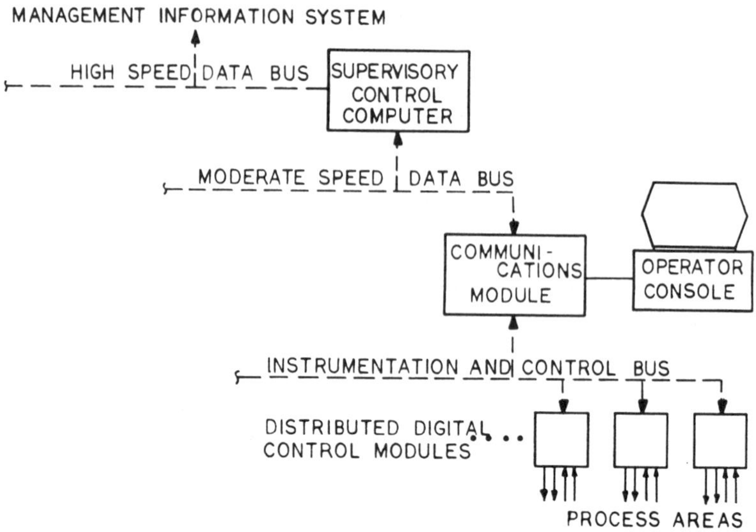

FIGURE 4.5 One alternative approach for configuring a distributed control system is where multiple local instrumentation and control busses are interfaced to higher level computers in a hierarchical configuration utilizing smart communication devices.

connected modules can be operated locally or from operator stations via the data network.

Process connected modules accommodate all input and output signals—both analog and digital—required to effectively monitor and control the process.

4.2.1.1 Basic Control Modules

The basic control module is the fundamental building-block for the entire distributed control system. All data acquisition, conversion, limit checking, and control take place in the basic control module. In addition, the basic control module also supports communication to the network and performs self-diagnostics.

The basic control modules are typical design to include the following hardware components.

Microprocessor

Read Only Memory (ROM) permanently maintains all master instructions necessary for control module operation, such as control algorithms, diagnostic routines, and executive programs.

Random Access Memory (RAM) maintains unique application data such as configuration words, setpoints, process variables, and tuning constants. RAM memory may be either backed up by core memory or provided with battery backup to maintain current process data in case of a power failure.

The most important characteristic of the microprocessor based controller is the software structure. These controllers are not programmed in the usual sense of the word. Instead each controller contains a full complement of selectable complex control strategies. These strategies reside in the ROM memory as control algorithms. Depending upon the particular controller, the number of algorithm available to the engineer ranges from 28 to nearly 100.

When developing a particular control strategy for a controller, the engineer selects the desired control algorithms and links them together. This process is typically referred to as configuration to differentiate the operation from programming.

Typical control algorithms performed by the basic controller are as follows:

P, PI, PD, and PID control
Lead/lag and dead time compensation
Nonlinear and ratio controls
Square root extraction and thermocouple linearization
Arithmetic calculations such as summer, multiplier
Logical operations, such as high select/low select

In addition to control algorithm calculations, all basic control modules provide functions such as process variable tracking, control initialization, etc. Users can configure this type of functions into his system.

The method of configuring a controller also varies considerably between vendors. Various alternatives are:

Data entry panel which can be shared by several controllers and which also provides a permanent operator display for each controller on a selectable basis (1).

Hand-held programmer which plugs into the controller being configured. Once the configuration has been completed, the programmer is removed. This method is typically used for controllers which have integrated faceplate displays (8).

Function button configuration panel which is built into the side of each controller. The engineer can, with the use of a very limited numerical display, proceed through a very set procedure and build a control strategy which includes all of the necessary control parameters (9).

Down load from a more sophisticated device over the data communications network. This is the typical mode for configuration of individual controllers which are a part of larger installations where operator consoles and bulk memory devices are available.

The capacity of a basic control module varies greatly from vendor to vendor. A module may consist of one to several functional blocks. Each functional block could be used to perform one algorithm, such as PID, multiplier, or high select. Basic modules with higher capacity have the advantage of implementing sophisticated control algorithms by linking several of these functional blocks together. However, from the application standpoint, the great majority of control loops are the single variable type which needs only one functional block. Thus, the high capacity control modules may offer only limited extra value for the user.

4.2.1.2 Enhanced Control Modules

Enhanced or advanced control modules take on a number of different forms, again depending upon the vendor. But typically enhanced control modules incorporate considerably increased control capability, in terms of increased input-output capability and the number of functional blocks available, which can be up to hundreds of blocks. The controller has the same library of control functions as those for the basic modules, but the controller will service a greater number of analog and digital input/output. As a result such control strategies as feedforward/feed-back control, coordinated control and other control strategies requiring multiple-input and multiple-output capability are more easily configured using a single enhanced control module. The enhanced controllers can also provide increased mathematical logic functions and capability to perform sequence control.

Mathematical functions The large numbers of available functional blocks in an enhanced control module permit advanced calculations at the controller level. This makes it feasible to control a process based on a calculated variable, even if such a calculation requires many steps of arithmetic operations. This also permits a greatly expanded variety of control functions to be incorporated into the control system design such as various predictive algorithms and adaptive control criteria.

Logical functions The addition of logical functions to the controller capability greatly enhances the potential applications. This function allows the controller to accept digital inputs, to use resident logic algorithms (such as AND, OR, and COMPARE), and can be configured by users to implement advanced control functions. Typical examples for these are nonlinear functions, such as control strategy selection, (select Equations A or B), controller modifications (select PI or P), input signal selection (select high, select alternative upon failure, etc.) and automatic initialization.

Digital/sequential batch functions One of the most recent and significant developments in distributed control system evolution is the addition of the sequence control function to the controller repertoire.

Sequence control greatly expands distributed control capability to include automatic start-up/shutdown of continuous processes, control of semicontinuous processes such as reactor switching and catalyst regeneration, heater decoking, soot blowing, and control of pure batch processes.

Additional microprocessor capability is provided which gives the ability to integrate continuous control loops and control functions, which have previously been handled by relays, timers, cams, and programmable controllers, into one control strategy. The sequence logic has the capability of turning on a continuous control loop at the proper time, initializing the loop with a preset setpoint value, and then modifying the setpoint value as a function of time, process condition, or some other predetermined profile, to achieve the desired end results.

In some cases a specialized higher level programming language has been developed for the sequence control function to facilitate ease of program development.

4.2.1.3 Data Acquisiton Modules

Operators and process engineers often need more than just controller related information to effectively manage their processes. Many additional analog and digital variables, such as uncontrolled flow rates and equipment status information are required. To help gather this information, several distributed control systems incorporate a distributed data acquisition system of intelligent microprocessor based units connected by the data communications network to the operator consoles.

Data acquisition modules are available in three varieties which scan high level input signals, and low energy signals such as thermocouples, resistance temperature devices (RTDs), and other millivolt transducers, and multilevel system servicing a variety of signal levels and types with a single processor.

The data acquisition modules provide:

Significant saving during installation by reducing long wiring runs.
Optimal system response to operator commands with data preprocessing.
Minimal maintenance requirements using self-diagnostics.
Reduced spare parts inventory based on circuit card commonality.

High level process signal interface module The high level process signal interface module acts as a high-speed analog input scanner, digital input/output interface, and analog output subsystem, and provides:

High level analog input data acquisition and processing.
Digital input data acquisition and monitoring.
Analog outputs for noncontrol functions, such as sending a signal to a conventional chart recorder.

Digital outputs for noncontrol functions such as an annunciator panel.

Important analog input preprocessing includes: high/low and deviation alarms, input square root extraction, and digital filtering that eliminates process noise.

A high level process signal interface module may have a capacity of handling hundreds of inputs and/or outputs. The major advantage for using this module is that it costs far less than other modules when the number of points are high. Note that one can use a controller module for the purpose of data acquisition and also be more economical when there are just a few points to be processed.

One could use this module as the process input/output device for a supervisory control computer, and to implement control functions, but this is not a good engineering practice because of the cost in software and the lack of hardware failure protection. It is better to perform control functions by using controller modules.

Low level process signal interface module The low level process signal interface module provides high speed, high accuracy scanning of low-level analog inputs from the thermocouples, resistance temperature devices and linear millivolt transducers. There are no functional differences between this module and the high level interface module.

4.2.2 Communications

The advent of distributed control system architecture resulted in the introduction of a new dimension into control system design concepts, that is, digital communications. The distribution of control hardware throughout the plant to provide maximum reliability and security, yet operating from a centralized operator's console, implies the necessity of designing a very secure, reliable, and efficient data communications capability. These goals are being accomplished through the use of asynchronous digital communications technologies. Data are being transmitted as bit-formatted messages, rather than voltage or current signals. Consequently, infinitely greater amounts of data can be communicated over a single pair of wires with greater security and reliability.

Data communications are generally accomplished over two-wired circuits, either as twisted pairs or coaxial cable, or by fiber optics. Beyond these two common characteristics, the variations between communication system designs from vendor to vendor are as numerous as the number of vendors. However, some additional functional aspects can be further discussed.

4.2.2.1 Communication System Configurations

The simplest distributed control system communication is the single multidrop coaxial cable (this is usually installed in a redundant configuration) upon which all control, data acquisition, and operator communication (console) devices reside. There exists a maximum number for the devices that can reside upon a single cable; usually the limit is the device address word length.

When additional devices are required, another cable is installed. Each cable usually requires a dedicated operator console.

This type of communication is typically operated in a polling mode where periodically a master, usually the operator's console microprocessor, would poll the devices on the cable to determine status and acquire desired data. Operator initiated changes are also communicated to the controllers via this cable.

No messages are initiated by the controllers and no communications exist between devices residing on the cable, other than the master. This system requires a traffic director to control the flow of communications to the devices on the cable.

The next level of design sophistication is a system which incorporates two or more levels of data communications, still using the two-wire cable. The first level of the multilevel design configuration usually provides the capability of intercontroller communication; thus, information can be carried between controllers allowing complex controller configurations to be developed. This level permits controllers to communicate with each other without a need for other devices, such as common electronics or traffic directors. This also permits the ends of the cable to be connected, forming a loop permitting communication in either direction, and, thus, increased redundancy. The data communication rate at this level is relatively slow, in the order of 100–200 K baud.

The second level of data communications interfaces several controller communication cables providing data flow to the central operator's console for a processing area. This level of communication is most likely operated in a polling mode and has a data communication rate in the order of 1 M baud.

The highest level of data communication offers very sophisticated data transmission capabilities providing the data conversion, buffering, and processing necessary for an efficient communication of information between devices or lower level cables which have different communication protocols and speeds.

This level of communication typically moves large amounts of plant data to and from high level computers performing sophisticated control functions, such as the implementation of integrated control strategies and plant optimization. Communications speeds at this level often exceed 5 M baud.

4.2.2.2 Communication Modes

As was previously mentioned, the performance of a communication network, regardless of the configuration, whether it be "star", "loop", or "hierarchical", greatly depends upon which of the following design criteria, have been selected—baseband or broadband, and active or passive (7).

Baseband/Broadband A design where data bit pulses of the serial message are placed directly on the line is referred to as baseband communications. As data communication speed increases, the width of the bits and the space between bits is necessarily decreased. Ultimately, line distortion and attenuation will limit transmission speed or distance using a standard coaxial cable. A rule-of-thumb states that at 1 M bits per second, the maximum distance will be 10–15 thousand feet.

However, baseband transmission is relatively simple and provides adequate reliability within that distance limitation and is, therefore, used as the basis for most process control communications networks.

Broadband transmission takes the data bits in a message and modulates them on a higher frequency carrier signal: attenuation and distortion are less of a problem and transmission at greater communication rates can be accomplished over many miles on a single coaxial cable.

Current technological advances in fiber optics, particularly in the area of eliminating the repeater stations, which are relatively slow and inefficient, and replacing them with external optical amplifiers are promising data communication speeds in the 100 M baud range over unlimited distances.

Active/Passive The means by which data are communicated from one device residing upon a communications network to another depend upon whether the network design is active or passive.

An active network design is one which requires a third piece of active logic hardware to be involved in the communication process between the two devices. Most process control data communication networks today are active in this sense. The third participant in the communication process, often referred to as the traffic director or bus master, assures orderly communication throughout the network by performing the following functions:

Establishes network device communication priority
Repeats messages for alternate network branches
Switches to backup network on command

Typically, the traffic director will be configured redundant, in conjunction with the redundant network cable, since its failure causes the entire communications system to fail.

The advantage of the active network design was that the devices attached to the network could be relatively simple in design and, therefore, less expensive. With continuing price/performance improvements in microprocessors, this advantage is becoming less important.

The passive network is designed eliminating the third device, traffic director or bus master, and providing all devices residing on the network with additional intelligence.

The disadvantage of the passive network design philosophy is that new logic must be provided to permit resolution of the contention problem as multiple devices attempt to access the communications network simultaneously. Various methods have been developed to resolve this problem and yet retain efficient use of the network. The key factor in any passive system design is that of guaranteeing that time critical data are communicated within a reasonable time frame on a heavily loaded bus.

4.2.2.3 Communications Security

In order to provide the high level of communication security that is required for a distributed control system extensive error checking procedures are incorporated into the communication system protocol. The types of error checking include such procedures as the following:

Repeat the message and compare the two
Use of standard message lengths
Various handshaking protocols requiring acknowledgments
Data echoing where the receiver retransmits the message to the sender for verification
Address diagnostics
Error detection codes where information about the message is embedded in the message itself

Most communication networks utilize these techniques in various combinations to achieve the necessary level of message transmission security.

4.2.3 Operator Console

4.2.3.1 CRT Displays

As the television has become the window of the world for the average family at home, the CRT display has now become the process operator's window to his process. Until just recently, process information was provided to the operator by various means, but always in fixed format and limited number (as single points of data). This was accomplished via indicator lights, gauges (various sight glasses, dials, and meters), recorder charts, circular or strip, and multipoint recorders to name the most common. It was the operators' responsibility to assimilate all of the available data from the process, interpret the implication of the data using his own personal understanding of the process operation and intuition, and take what corrective action was necessary.

As control rooms became more sophisticated and controller hardware became miniaturized, single control rooms associated with individual process areas were grouped together to form a centralized control room; the motivation was to improve operator communication between associated process areas. This often resulted in control panels several hundreds of feet long and, consequently, the operators at times had to visually scan vast areas of the control panel to determine the status of a particular process area.

Also inherent in the design of the analog controller control panel was the problem of identifying large number of controllers and indicators with the associated process areas. In an attempt to provide a more meaningful display for the operator, the graphic panel was created. Graphic panels consisted of large panels containing colored cutouts representing major process equipment and associated major process stream piping. The gauges representing process data for each piece of process equipment, such as a heater, distillation column, reactor, or a heat exchanger were located on or near to the cutout representing the equipment on or adjacent to the inlet or outlet piping. Thus, the flow rate and temperature for the charge stream to a furnace were shown located in close proximity to the piping cutout representing that flow.

It was found that as the number of data points increased, the physical size of the graphic panel become untenable. The outcome was the semigraphic panel. An example of this style is shown in Fig. 4.6. The instruments and gauges were tightly packed on the lower portion of the panel to minimize board space, but still remain accessible to the operator. The graphic representation was then made much smaller and usually placed above the instrumentation area. The graphic panels then had printed tag names at the key process areas which were associated by the operator to the tag names on the gauges and recorder located in the lower panel.

Both of these types of operator display panels had a common deficiency. This was inflexibility or rigidity. As is the nature of all processes, they change. Improvements or product specification changes often result in changes in the process configuration. It is very difficult, if not impossible, to reflect these changes on the rigid graphic panels, so sooner or later they become out-of-date, they do not represent the current process, they no longer fulfill the function for which they were designed. Consequently, they represent an expensive room consuming antique.

Engineers have made many designs and improvements on computer driven displays since the first process control computer was installed. Some CRT displays and special keyboard data entry devices were implemented with computers as early as the late 1960s. Those devices were very expensive and they were applied only by users who needed

FIGURE 4.6 An early approach to process graphics was the semi-graphic panel which displayed the process flow, but located the instrumentation separately.

sophisticated information displays to run special processes. Experiences from these early day installations have helped to make CRT the standard operator communications devices for distributed control systems.

Several important facts about CRT displays need to be discussed:

"Human engineering" is extremely important for the design of CRT displays and data entry keyboards. For example, the use of the standard typewriter keys is not considered a very good practice, but to make an operator keyboard arrangement can be a major human engineering task. Similar efforts are also needed for designing displays.

It takes great "software engineering" or programming effort to implement operator communication functions, thus, users should not try to develop their own systems. (Users may "generate" their own displays from vendor supplied software packages. There is a great difference between generating a display and developing a new system.)

Because of these characteristics, every distributed control system vendor offers a very different design in CRT packages from all the other vendors. Since human engineering can be subjective, everyone claims that his design is the best, and it is not an easy task to compare and evaluate these systems. The only sure answer for CRT systems is that each system has some advantages over the others, but no one system can be said to be the best. The following discussion lists the potential functions which can be performed by a system, but does not imply that all of these are provided by vendors as standard items.

The CRT operator display is a panacea for the operator in that it removes most of the difficulties associated with the conventional control panel. First and most important, the CRT display presents information to the operator about the process, not just data. Such measures of process performance as severity, conversion, yield, profit factors, etc., can be displayed upon request. No longer need these be based upon operator's intuition.

Secondly, the single CRT screen, or several screens side by side, can present information on many processing areas in response to the pressing of a key. No longer is the operator required to course up and down great distances of control panel to determine the statuses of related process areas.

Thirdly, the CRT screen can be used to display current trends of a number of related process variables on an exact time scale or to recall and display historical data again on an exact time scale. No longer is the practice of unrolling strip charts necessary to determine past trends of a single variable, or even more difficult, the correlation of several strip charts to determine process interactions. It is nearly impossible to determine the exact time frame for each chart. Archiving strip charts and the subsequent frustrating task of locating the exact charts associated with some past event to review the circumstances is no longer necessary. Exact process data with accurate time references can be easily and inexpensively stored on magnetic disks for easy referencing at some future time.

Finally, the CRT screen offers graphic displays, and they are flexible. If the process changes, if new measurements or data points are added, if the operator or process engineer desires an additional display of a particular area, all of these can be simply and easily accommodated through the configuration or modification of displays via the function keyboard. The displays will always be current. The flexibility of CRT systems makes it possible for engineers to specify operation-oriented displays at a low cost. Note that a conventional panel mounted instrument, such as an indicator, can only be located at one fixed location. It can be part of only one particular group display, unless someone is willing to pay the cost of more indicators for the same variable. In real plant operation, a process variable may need to be shown in several group displays to help the operators. For example, fuel flows to a heater should be part of the heater temperature control display, and also be part of the plant fuel balance display. These types of displays can be easily arranged using a CRT based system.

There are five typical types of CRT displays. These represent information about the process in the following forms:

Analog graphics Analog graphic representation of process variables relative to desired values. This form of display is generally used in the overview which is designed for quick operating surveys of a large area of the process or processes allowing an operator to see at a glance the deviation of each loop from the desired value. This display is comprised of groups of vertical bars, one for each analog input, where each group of bars, typically 8 to 16, is associated with a small processing area such as a column or a heater.

The bars are color coded such that when the deviation exceeds an alarm limit, part or all of the bar changes color to indicate the severity of the alarm. The bar may normally be green. Yellow would indicate a low level alarm requiring operator attention where red would indicate an emergency alarm requiring immediate operator attention.

Numerical displays Very frequently, displays will be comprised of numerical data intermixed with alphanumeric tag names and engineering units. Rather than presenting the operator with vast arrays of numerical data which just fill the page, but have very little association with the process and, therefore, are difficult to comprehend, the data are often grouped in meaningful formats.

One format, which is a carry-over from the familiarity with analog instrumentation, is the faceplate display which presents a pseudo-representation of the analog controller faceplate using a combination of alphanumeric and analog graphics. Usually, several of these faceplates are grouped in a single display again relating to a small process area.

The information in one faceplate can include:

Name of the point, loop, group, or device
Descriptor of the point, loop, group, or device
Numeric and graphic measurement of an analog point or loop
Contact state of a digital point or group
Current command to a digital device
Color-coded alarm type and status for a point, loop, group, or device:
- Blank for no alarm
- Yellow for non-critical alarm
- Red for critical alarm

Control status (automatic or manual) of a loop
Cascade status (closed or open) of a loop
Operational status of a device

Numeric and graphic setpoint and output of a loop

The faceplate display permits the operator to see the control actions required. He can then perform the action and watch the effects. Control actions supported by the faceplate display include:

Alarm acknowledgment
Numeric and ramped loop setpoint modification
Numeric and ramped loop output modification
Loop control status modification
Loop cascade status modification
Digital point and group contact state modification
Digital device command entry

Alphanumeric information can also be used to create a more detailed single-loop display.

The single-loop display provides maximum information for one loop, point, or device. The operator process engineer is presented with a faceplate and a listing of all limits and tuning parameters. The display permits performance evaluation and updating of all operator control functions, as well as loop tuning functions, such as altering a proportional gain constant.

With the single-loop display present on the screen, the operator/ process engineer can typically perform all of those functions supported by the faceplate display as listed above, plus the following:

Change point or loop alarm limits
Change loop tuning parameters

Single-loop display is only needed during setup or troubleshooting a control loop.

Graphics displays Interactive graphic displays are one of the most significant features of the CRT console. The graphic display does not only present a pictorial representation of the process equipment and flow diagram (with valves, pumps and other equipment shown), it also is capable of showing the control system configuration.

Furthermore, the screen can be overlayed with live process data showing current values of process variables, controller setpoints, controller output, plus loop and equipment status (valves open or closed, motors started or stopped). An additional capability is to be able to display calculated data on the process graphic displays (and laboratory inputs).

In most cases, the operator can directly access information on the screen either via a cursor or in the case of touch sensitive scans, by touching the desired data point. He can directly change setpoints, limits, loop status (auto, manual, cascade, computer), and equipment status (on/off, open/close, etc.). The use of changing colors and flashing characters is very convenient for emphasizing change of state of a particular variable.

Trend displays The current trend display provides the operator with a graphic comparison of the real-time measurements for from two to usually several data points.

The trend displays program usually allows several alternatives when building the display. The limits are selectable so that the trend is within a reasonable range. The data may be selected as current value or some smoothed or averaged value of the current values. The update time is also selectable and ranges from a point per scan when evaluating controller response during the tuning process to once per several minutes when observing long-term trends.

Historical trends are similar to trend displays except that they usually capture data at a less frequent interval, the data are usually saved for 24 hours or longer, and the data are saved on a bulk memory device, such as a floppy or Winchester type disc. Historical data are usually retrieved and displayed for analysis at a future time using the host driven operator console.

Text It is very convenient to provide operator emergency instructions triggered by a shutdown alarm or even normal instructions for a sequence type start-up/shutdown procedure that does not occur often and the operator's memory may be rusty. The particular instructions desired can either be requested or can be automatically displayed as a result of some process condition.

The instructions can be interactive in a sense. That is, the text may instruct the operator to perform a task not under the control of the instrumentation system, but the system can monitor the process to determine that the instruction is being executed correctly. For example, the text may instruct the operator to fill a vessel with water and wait 15 minutes before proceeding to the next step which may be turn on the agitator. The instrumentation system will monitor the tank level to verify that it was filled. If the agitator is turned on too soon, or not on time, the system can alarm this failure, etc.

4.2.3.2 Operator Input

Keyboard The operator console is the operator's gateway to the process. All of his commands and requests must be entered via the operator keyboard or some functionally equivalent device. Therefore, the functional design of the operator interface is of the utmost importance. The function must be easy to learn and remember. It must be designed to enter information with the minimum of keystrokes, with security and without confusion. Misdirections and spurious inputs must be trapped and either ignored or alarmed. Input data must be checked for reasonableness.

The evolution of the operator console keyboard has followed several divergent paths. However, it can be stated that the primary goal of each design alternative was to simplify the opeator's task of entering data and receiving information and increasing the security of the system by eliminating erroneous inputs.

The trend has been toward fewer keys with increasing use of function buttons, one keystroke completes an entire function, multifunction buttons where the function of the button changes with the display, and various touch-type displays. The design of the operator console keyboard has moved far beyond the early simple typewriter keyboard, these early keyboard designs proving to be slow, cumbersome, and subject to frequent operator errors.

Variable function keyboard: Several consoles are designed using both fixed and variable function keys. The fixed functions (labeled) keys select various types of operations, such as alarm handling, trending recording, system configuration, etc. The variable functions keys are assigned different functions, depending on the current display. The current functions associated with each variable function key is displayed at the bottom of the corresponding CRT display directly over the button. In many cases, the fixed function keys are not truly fixed in that their function is programmable and can be assigned at the time of system configuration. However, once a function is assigned to a "fixed" key, it cannot be changed without reconfiguring.

Dynamic touch display: The dynamic touch display is an interactive keypad whose "buttons" (touch sensitive areas) are coordinated with the particular display on the operation display screen. The purpose of the pseudo-keypad is to indicate to the operator his next set of display options. Touch a label to select a new main display and the dynamic display screen updates with your next series of choices. After calling up an overall plant layout on the main screen, the dynamic display would provide labels for all the main process areas. Touch one of these, and the main screen changes to a close-up of that process area. Usually a relatively few function buttons are associated with the dynamic display to permit entering control setpoints, tuning parameters, and changing loop status.

Touch screens Touch sensitive CRT screens for changing display modes used in conjunction with a data entry keypad is another approach to minimizing operator access time to process information. A complete hierarchy of displays is instantly available at the operator's fingertips. An overview display may give the operator the current status of the entire plant at a glance. Merely touching one of the group in the overview will call up a specific processing area either as a graphic display showing current values of operating data, process variables, controller statuses, equipment status or as a faceplate display, which presents a simulated controller faceplate for each control point in a particular group, usually 8 or 16 points. Touching one particular faceplate selects that point for operator action. Another touch will call up a highly detailed single loop display. The operator can review all of the information related to that particular control point and modify the controller parameters using the associated display keypad.

Voice commands An operator interface system is currently under development which is designed to respond to a series of voice commands which have been previously selected. Even with its limited vocabulary, this type of system will be extremely functional, eliminating the many keystrokes normally associated with information retrieval. The system is built by assigning key words or phrases to particular functions. The computer then listens for these key phrase commands and when one is detected, responds in an appropriate manner. The computer is calibrated to each individual operator's speech characteristics by digitally recording several key phrases for future reference.

4.2.3.3 Data Storage/Retrieval

The historical data storage/retrieval function serves as system storage for process history, schematic displays, and system records. Operation of this activity is usually automatic and with the exception of floppy disk archiving, completely transparent to the operator.

Current trends in distributed control system designs include one or two Winchester disk units for storing process data. This reduced the frequency of archiving data on to an off-line system.

Information stored by the historical data storage/retrieval functions may include:

Continuous process history
- One minute samples for assigned points plus
- six-minute, hourly, shift, daily, and monthy averages for assigned points

Batch process history
- Receives and stores the history from an application module.

Process event history
- Event history for each process unit can be stored. Operator actions and diagnostic events are also stored.

Schematic displays
- Including help displays, are stored for each console area.

4.2.4 Configuration Function for Digital Controllers

4.2.4.1 Functional Description

Most digital controllers are designed to be sufficiently flexible to accommodate the requirements specified by the user. Earlier versions of digital controllers were designed to imitate single variable analog control functions, such as proportional-integral-derivative algorithms, cascade, or ratio control. Functionally, a control loop can be subdivided into the following five major blocks as indicate in Fig. 4.7. The hardware structure for implementing these functional blocks can be simplified as shown in Fig. 4.8.

Note that in Fig. 4.8, multiple inputs and outputs are connected to the controller. For the earlier version digital controllers, this was done for reasons of economy due to of the high cost of microprocessors at that time. Inside the digital controller, each input and output is treated as a separate loop. Users can connect two or more loops together within the same digital controller by software, but each loop has only limited functional capability similar to that for an analog controller. Contemporary digital controllers are designed with greatly expanded memory and CPU capacity, thus, they can be used to implement more sophisticated functions. These controllers can control anywhere from one loop configuration to complex control strategies with a large number of loops.

The most important advantage for the one loop per controller configuration is the distribution of risk; and the most important advantage for multiple loops per controller configuration is the ease of internal communication. There are some single loop controllers which are designed to communicate with other digital controllers, thus, the designers claim these to have all the advantages. However, some of the cross-controller communications are very slow; therefore, a control function requiring frequent communication between two loops, such as between

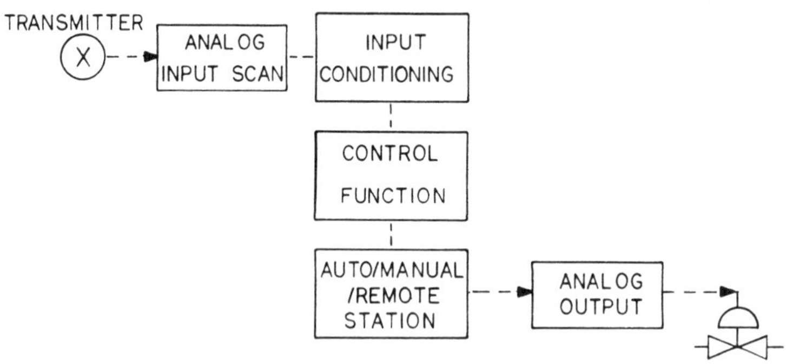

FIGURE 4.7 The single-loop controller can be defined as five function blocks which include: analog signal from flow transmitter input scan, input signal conditioning, such as filtering and limit checking, control function, an auto/manual/remote station to permit the operator to select the control mode, and an analog output signal generator to the control device.

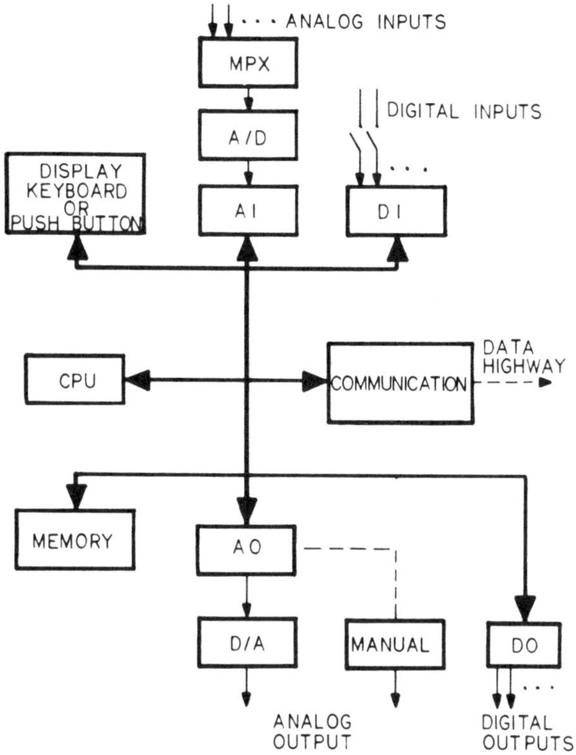

FIGURE 4.8 The hardware structure of a microprocessor based digital control loop consists of numerous function blocks. These facilitate the processing of analog and digital input signals from the process, operator input request, calculations, communications, and the generation of analog and digital output signals to the process and operator displays. (MPX, multiplexer; A/D, analog-to-digital converter; D/A digital-to-analog converter; DO, digital output; AO, analog output; AI, analog input.)

cascade outer and inner loops, must be implemented by hard wiring the two together. This results in a less flexible control system because any changes in control functions may require wiring changes between controllers.

Digital controllers are not programmed per se, but must be "configured" by users. The purpose of the configuration procedure is to permit the setup of a controller by an engineer without his having to know detailed programming techniques. The advantage of this approach is that customs controllers can be setup from a given set of standard hardware and software modules in simple manner.

The configuration tasks usually include the specifications of all control loop parameters as follows:

Analog input scan and conditioning Most digital controllers are designed to accept analog signals in the range of 1 to 5 volts or 4 to 20 ma dc. Some can also read thermocuples or (RTDs), but most systems don't. The analog input signals are multiplexed and converted to a digital form through an analog to digital converter; the user must specify the address for each loop input. The scanning frequency of the input signal is usually fixed according to a fixed sampling interval, such as a cycle every 0.25, 0.5, or 1.0 seconds. The users cannot change this sampling frequency, but a digital filter can be specified to filter out high frequency noise.

The raw data from the A/D conveter is usually checked with high/low limits for instrument errors, converted to engineering units and then checked with another set of high/low limits which are used for process alarms. The process high/low limits are selected by the user. The engineering unit conversion processing can involve specialized linear conversions or nonlinear functions, such as square root extraction or polynomial function generation.

Control function processing There are numerous basic function blocks which are preprogrammed by the vendor for the user to specify during the system configuration stage. In general, these fall into seven major categories:

Arithmetic operations including: addition, subtraction, multiplication, division, absolute value, defining constants or variables, matrix addition, or matrix multiplication.

Functional relationships including: square root extraction, nonlinear polynomial or piecewise linearization function generator, trigonometric function, or logarithmic function.

Logic operation including: logical AND, OR, exclusive OR, NOT, GT (greater than), LT (less than), GE (greater or equal), LE (less or equal), EQ (equal) etc. These are used for Boolean operation or signal comparison and selection.

Selectors including: multi-input selection for maximum value, minimum value, average value etc.

Limiters including: rate limit for high/low or high/low saturation nonlinearity for safety purposes.

Dynamic functions including: timer, counter, totalizer, lead/lag function, differentiation, unit delay.

Control function including: PID algorithm, Smith-predicator algorithm, or PID algorithm with variations. Special user programmable algorithms.

As discussed elsewhere, each digital controller may have a capacity for processing one function to several hundred functions. The users can select the types desired and link them together to perform a required control function.

Auto/Manual/Remote station and output calculation When a digital controller is used to implement a simple feedback control loop, a manual/auto station is included as part of the controller. The user may specify a bumpless transfer mode for switching from one control mode to another. For this type of control loop, output is usually converted to percent of full scale or position output.

When cascade loops are implemented, an Auto/Manual/Remote station is included. Figure 4.9 shows a cascade loop. Achieving bumpless transfer for this loop is much more complicated than for the simple loops, but most digital controllers provide the capability for implementing bumpless transfer by selecting one flag or by linking a few function blocks together.

A digital controller must be configured according to the method developed by its designer. It is not practical to discuss all of the configuration methods because there are just too many vendors in the market now, and more new vendors are coming. The following examples describe the configuration procedures for three widely used controllers.

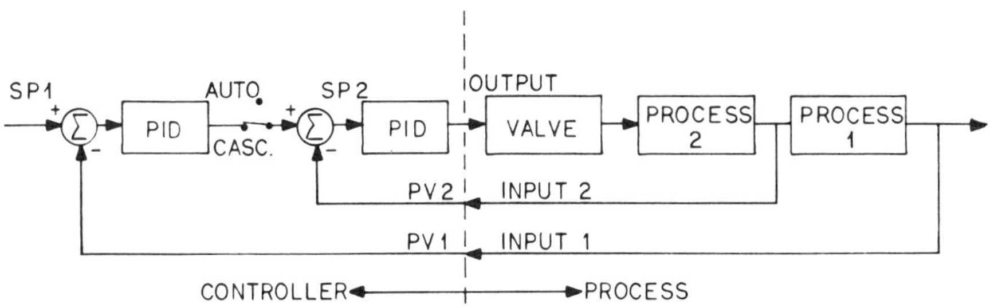

FIGURE 4.9 The configuration of a cascade control loop consists of an inner loop, SP2 and Process 2; and an outer loop, SP1 and Process 1. The operator determines the setpoint for the outer loop. A deviation between the measured value of the controller variable, PV1, and the setpoint SP1, will result in an adjustment on the SP2. A deviation between the measured value of the controlled variable, PV2, and the setpoint, SP2 will result in a change to the output to the valve.

4.2.4.2 Configuration of Simple Digital Controllers

Honeywell's TDC-2000 basic controller was the first of its kind. It has limited capacity, but is also easy to configure. Each controller can accept 16 analog or digital contact inputs and service eight analog or digital outputs. It is not economical to use this controller solely to scan digital contacts. The major function of this controller is to replace analog controllers. A basic controller can be used to replace eight analog control functions, such as a feedback control loop, a ratio loop, a selector, etc; (1).

A specific control loop can be set up by specifying four configuration words. Each word is a four-digit number. Readers may consult a Honeywell manual to investigate the details of configuring a loop. Basically, this is a simple procedure and a user does not have to be familiar with the principles of computer programming to learn this technique.

An example of how a flow controller, shown in Fig. 4.10, can be configured follows: (Slot No. 7 of a controller is used.)

Word A	0 1 0 3	
Where	0 1	Normal PID algorithm
	0	Past mode recal in manual
	3	PV tracking and initialization
Word B	7 1 7 3	
Where	7	x input address
	1	PV type signal
	7	y input address
	3	sp type signal
Word C	2 0 0 0	
Where	2	Decimal point location xx.xx
	0	Linear display
	0	Not applicable
	0	Deviation alarm
Word D	0 0 1 1	
Where	0	"A" type equation for PID
	0	Direct output indication

1	Reverse control acting
1	Square root input conditioning

These four words must be specified to set up this control loop. Many other variables also have to be specified, such as engineering unit codes, filtering constants, limits, etc. These variables can be either specified during loop configuration, or after a system is generated.

The advantages of this type of controller are that they offer all of the advantages of digital controllers as discussed elsewhere, and they are very easy to apply. There is, however, one major shortcoming for this type of controller, that is limited capacity. The cost of implementing a sophisticated control function using this type of controller is high. Note that each controller provides eight functional slots, and each computation function, such as an adder, will utilize a slot. A flow control loop based on temperature and pressure corrected flow rate will typically require five slots, and a more complicated calculation may require several controllers.

The high cost for implementing advanced functions may not be as bad as it sounds. A process plant (based on continuous process) usually contains far more simple control loops than advanced control loops, so the incremental cost for implementing advanced functions could be small as compared to the total cost.

4.2.4.3 Configuration of an Advanced Digital Controller (1)

There are many advanced digital controllers offered by vendors. Most of these controllers can accept one to several analog or digital inputs and generate one or two analog or digital outputs. The major purpose for this type of controller is to implement advanced control functions, so each controller provides a capacity for many functional blocks (200 to 300). The control system design concept is based upon using one controller to implement one sophisticated control algorithm which requires multiple computations by linking multiple functional blocks together.

Users do not need computer programming experience, but the effort and knowledge required to configure such a controller is more demanding

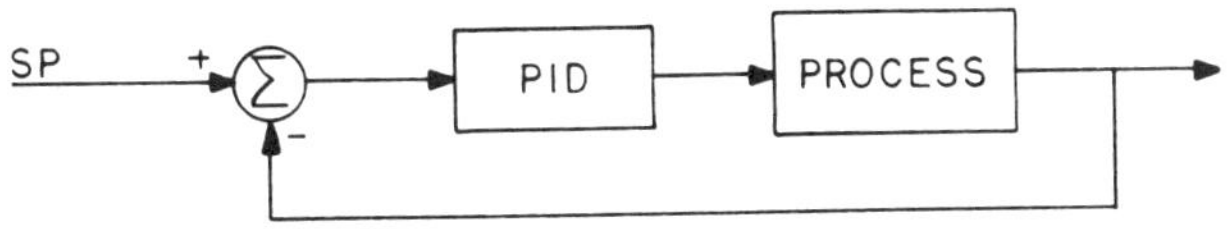

FIGURE 4.10 Functional diagram of a flow controller with a proportional-integral-derivative control algorithm.

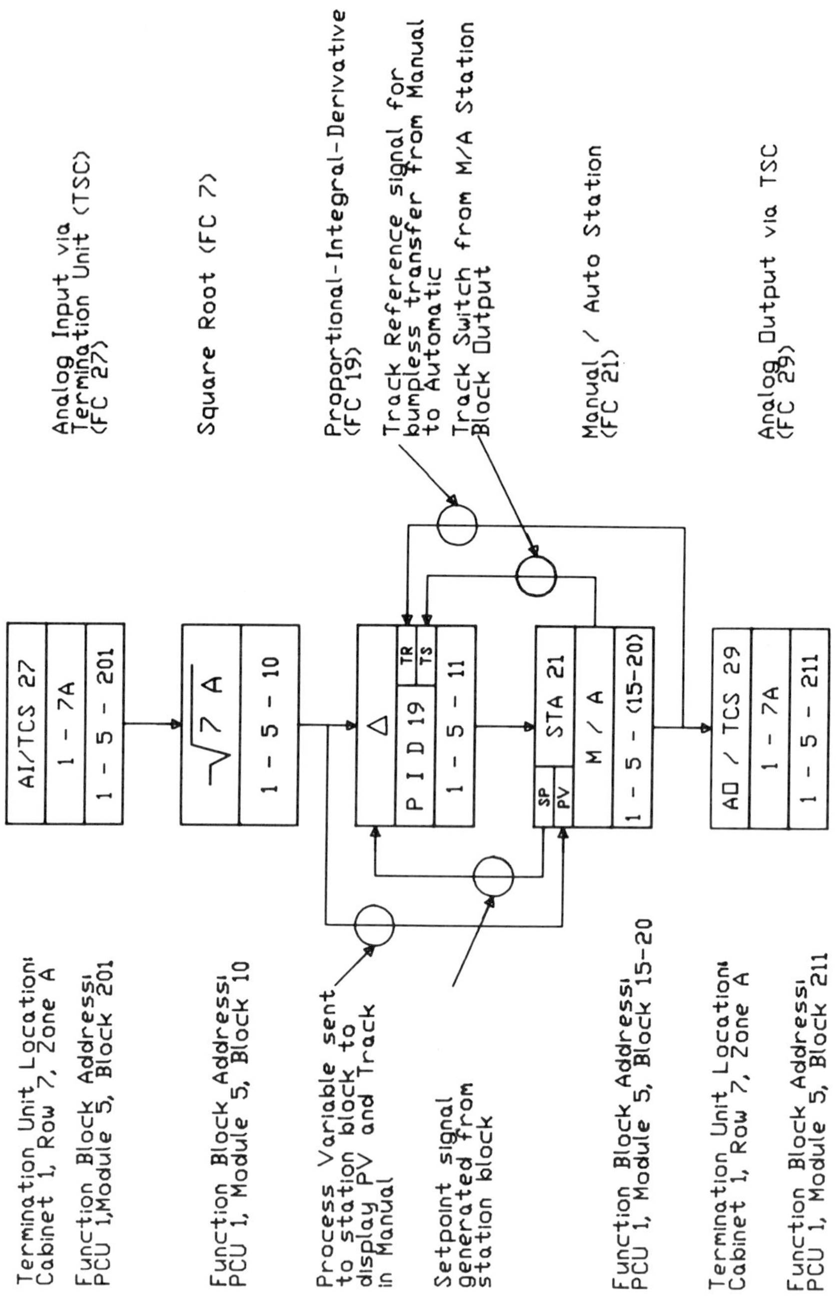
Termination Unit Location: Cabinet 1, Row 7, Zone A
Function Block Address: PCU 1,Module 5, Block 201
AI/TCS 27
1 - 7A
1 - 5 - 201
Analog Input via Termination Unit (TSC) (FC 27)
Function Block Address: PCU 1, Module 5, Block 10
√7 A
1 - 5 - 10
Square Root (FC 7)
Process Variable sent to station block to display PV and Track in Manual
Setpoint signal generated from station block
Δ
P I D 19
TR
TS
1 - 5 - 11
Proportional-Integral-Derivative (FC 19)
Track Reference signal for bumpless transfer from Manual to Automatic
Track Switch from M/A Station Block Output
Function Block Address: PCU 1, Module 5, Block 15-20
SP
PV
STA 21
M / A
1 - 5 - (15-20)
Manual / Auto Station (FC 21)
Termination Unit Location: Cabinet 1, Row 7, Zone A
Function Block Address: PCU 1, Module 5, Block 211
AO / TCS 29
1 - 7A
1 - 5 - 211
Analog Output via TSC (FC 29)

than that required to configure the simple digital controller previously described. Figure 4.11 shows the block diagram for configuring a simple feedback control loop as shown in Fig. 4.10, using a Bailey Network 90 controller. As shown in Fig. 4.11, a user needs to define five function blocks and link them together to accomplish the same control function which can be defined in one block using a Honeywell basic controller. Again, readers may consult Bailey's users manuals to investigate the detailed configuration procedures. The following descriptions are made just to provide some broad ideas on the amount of effort required to generate a loop (10).

The method for linking the blocks together is to define, in each selected function block, the source address of the external function block for input. For example, the input source for block address No. 10 (the square root function) is from block address No. 201. For a function block where the multiple external sources are needed, all of the sources must be specified. For example, the function block address No. 11 (the PID function) requires four external input signals, so the user needs to specify the locations for all source addresses. The four required inputs are: (1) Process Variable, (2) Setpoint, (3) Tracking Signal, and (4) Tracking Switch.

The input specification is shown below:

Input		Source Address
1 Process Variable	=	Block Address No. 10
2 Setpoint	=	Block Address No. 16
3 Tracking Signal	=	Block Address No. 15
4 Tracking Switch	=	Block Address No. 17

The block addresses No. 15 through No. 20 are all located in the function block called Manual/Auto station block which is designated with the function code, 21.

Obviously, the configuration procedure for this controller is more difficult than that for configuring a simple control loop, such as a Honeywell basic controller. However, this type of controller is far more flexible than a simple digital controller, and can be used to implement much more complicated control functions.

An example given here is to use a single controller to simulate a first order process and a control loop. That is, the controller

---

FIGURE 4.11 Function block diagram for configuring a simple control loop on the Bailey Network 90 digital controller. (Courtesy of Bailey Controls Company, Wickliffe, Ohio, 1980.)

and the simulated model are combined in a closed loop manner using a single controller to accomplish the simulation. The block diagram of the control loop is shown in Figure 4.12. The configuration block diagram from this example is shown in Figure 4.13.

4.2.4.4 Configuration of an Advanced Digital Controller (2)

The configuration procedure for another group of advanced digital controllers is similar to the computer programming approach using a high level language. This configuration procedure is to accomplish the necessary arithmetic and logic computation for a control loop, utilizing the preprogrammed function modules (blocks). The major difference between these types of controllers and the group described in the last section is that these function blocks are more flexible than those described before, and the controllers are generally more capable.

Basically, the organization of the computation procedure is arranged as shown in Fig. 4.14. An example of this type of controller is the direct digital loop controller (DDLC) supplied by Toshiba. Several models of different capacities are being marketed. Like the Bailey controller, each controller can have many modules. Each function module is structured with one or two signal inputs and one signal output with several parameter inputs as shown in Fig. 4.15. By interconnecting the function modules in sequence, the desired control strategy can be implemented easily as shown in Eq. (4.1).

$$y = f(x_1, x_2, P_1, P_2, \ldots) \tag{4.1}$$

A simple flow ratio control is given in Fig. 4.16 and implemented with function modules as shown in Fig. 4.17. Note that the PID block has an incremental output, and the output tracking (for bumpless transfer when switching from manual mode to auto mode) is performed in the output block. Comparing this PID block and the PID block in the Bailey controller, it can be seen that the Toshiba PID block only performs the mathematical calculation for the algorithm while Bailey's PID block also performs the logic function of tracking. Bailey's PID block is more capable, but less flexible. Toshiba's system is flexible, but also requires engineers to configure the system effectively.

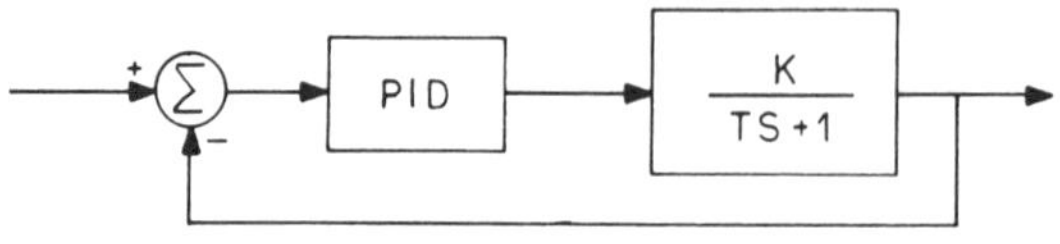

FIGURE 4.12 Closed loop control of first order system simulated using a single controller.

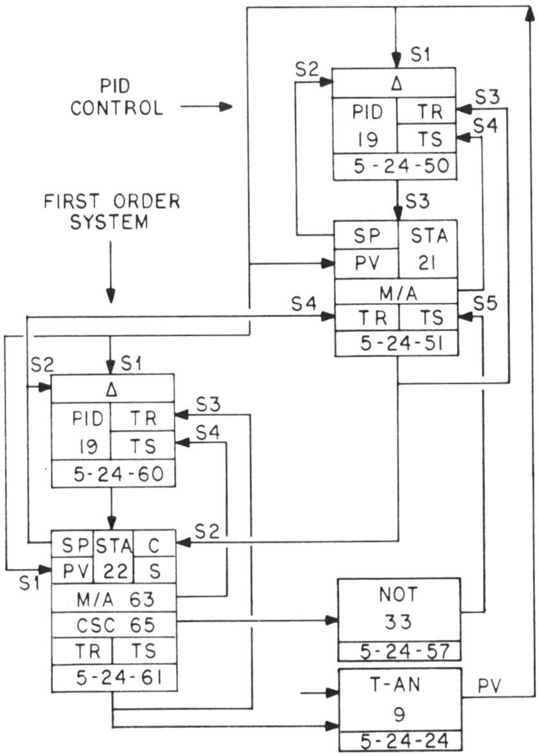

FIGURE 4.13 Actual implementation of closed loop simulation to illustrate using Bailey's Network 90 high performance controller module (COM03). This configuration simulates the PID control and first order process connected to form a closed loop system. The first order system is simulated with one PID function block and one auto/manual station block connected with feedback.

4.2.5 Host Computer

In the application of distributed control technology, one of the most important contributions is the improvement in safety and control system availability for the overall control of the plant. If the control system is designed utilizing the full potential of distributed control, a single failure will not affect the overall control of the entire plant. A failure in an individual controller will only affect the local control, and maybe not then, if backup redundancy is practiced.

The improvement in the overall safety is attributed to the design consideration of risk distribution accomplished by the distributed control system. However, for information collection and report generation

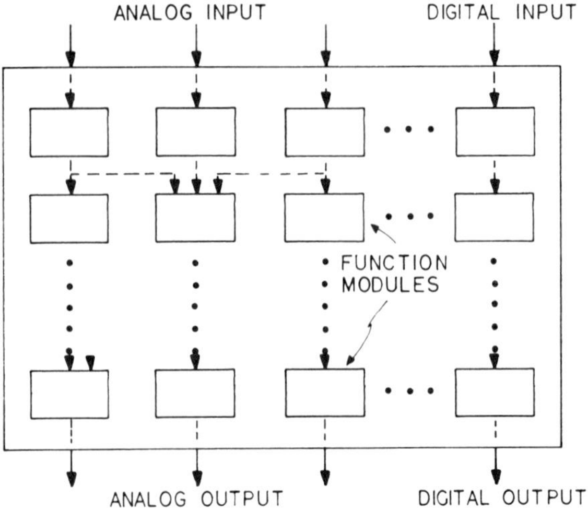

FIGURE 4.14 Computation procedure of a control algorithm created by interconnecting the function modules is illustrated. The function modules are preprogrammed algorithms with proper assignment for initial conditions or parameter selections. The sophisticated control algorithm can be created easily with this type of computation procedure.

for the entire plant, centralized information gathering and processing activities are required. These data processing activities are usually performed by a host computer residing on the data communications network. The size of the computer must be sufficient to accommodate the plant-wide information volume and real-time response capabilities. Since the host computer resides on the data highway, it has access to

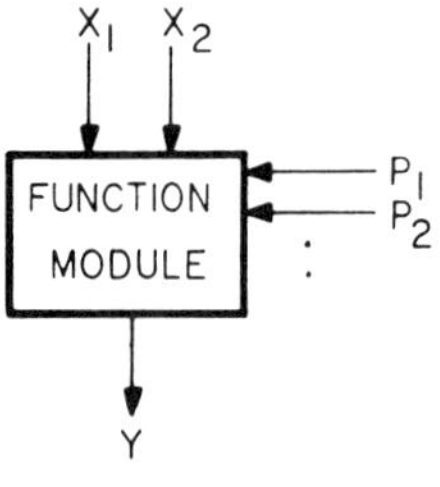

FIGURE 4.15 A typical function block is organized with multi-input and one output with constant parameters specified by the user for the block. The input X_1 or X_2 can be the output from other function blocks or external signals.

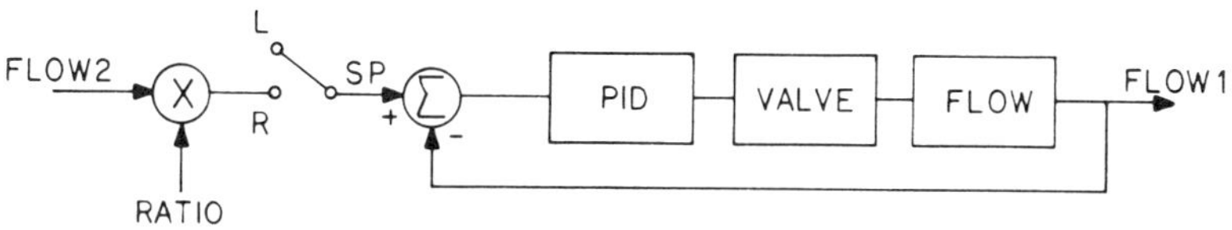

FIGURE 4.16 Functional diagram of flow ratio controller indicating how setpoint of Flow 1 is adjusted to maintain Flow 2 ratio to Flow 1 at desired value.

the data in the distributed controller and process input modules. Because of this indirect access to the process data by the host computer, the effect of the computer failure on the process control activities is isolated. This implies that a failure in the host computer will not affect the basic overall control of the plant using the distributed control systems. The failure of the host computer merely means termination

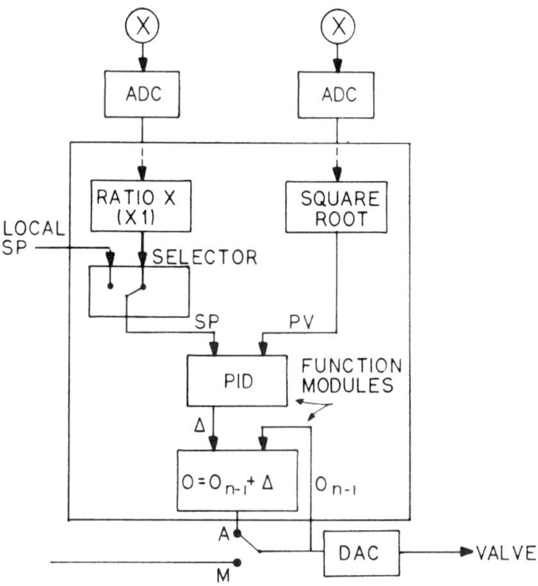

FIGURE 4.17 Actual implementation of ratio control of flow loop indicated by Fig. 4.16 is illustrated. The set point of the inner loop is generated from the output of a selector function block. One input to the selector is local setpoint and another input is from the output of the ratio block. The output from the PID function block is the incremental value, which is used to generate the output value to the valve.

of the centralized information gathering and supervisory control activities without endangering the overall basic control of the processes.

Host computers are usually able to perform supervisory control by sending appropriate commands to change the setpoint. In addition, the distributed controller configurations or tuning parameters or process high/low limits can be modified through the computer. In this case, adaptive control can be conveniently implemented. Since the task of performance identification is more suitable for a host type computer, not all of the host computers have the software to perform either adaptive, computer supervisory, or DDC control. A well-designed software system specification for process control is a major consideration for host computer selection.

4.2.5.1 Control Loop Processing

The host computer processes the process variables such as temperature, pressure, flow rate, analyzer input, digital input, etc., according to a specified update frequency which is determined by input computation, control action calculations, reporting, and alarming.

Process variables created in the data base are organized similar to those in the data base for a control loop. Therefore, the image of the controller parameter settings and limits can be transferred from the computer to controllers or vice versa. For each process variable a record or file is created using a name for identification. The character length should be sufficiently long so that the location of the instrument can be identified clearly. For a large petrochemical complex, the identification level may start from the region of the plant complex, process unit, instrument type, and instrument number.

For computer control of the process, either computer commands the setpoints of the controllers (this is called supervisory control) or the output of the controller (this is called Direct Digital Control), three major processing phases are required for the host computer.

Input processing This task performs the engineering conversion, limit checking, linearization, filtering, signal modification or selection, status checking, or alarm. The primary purpose of the input processing is to ensure that the data used for control or computation are accurate and reliable.

Control processing This processing phase includes the control computation for supervisory or DDC type control, adaptive tuning, logic computation, and multivariable control commands. The primary purpose of this processing phase is to generate a meaningful command for the manipulatable input to the process.

Output processing This processing phase includes output logic, setpoint magnitude or setpoint increment, analog output, pulse output, digital output, output limit check etc., so that the control command can be smoothly interfaced with the instrument or controller. The scaling of the engineering unit output must reflect the limitations of the physical requirement for the interface equipment.

4.2.5.2 Operator Support Functions

The host computer usually supports graphic generation and displays for operators or engineers for on-line operation.

Fault analysis Many system diagnostic functions, such as a data communications network and controller failures, can be indicated on the computer console to inform the operator and maintenance personnel of the problem area where the failure has occurred. The host computer is one of the most valuable tools for pinpointing trouble spots in the complex system. The safety of the entire complex of the petrochemical process can be improved substantially due to the instantaneous diagnostics provided by the fault analysis and alarming function of the host computer.

Management information For management information of the plant, reports, such as yield accounting of the plant and product inventory, can be generated periodically upon request. The historical data needed for report generation can be accessed through the data base of the real-time computer system. Periodically, the historical data may be stored on tapes or disks for permanent storage, or transferred to another computer for off-line processing for automatic data entry to the yield accounting programs or application programs.

A communications link between host computers for distributed control and the large data processing computer is sometimes required for a large plant to speed up the process of plant data update and generation of reports to the manager in time.

4.3 PROGRAMMABLE LOGIC CONTROLLERS

To explain the functional difference between a basic microprocesser controller and a programmable logic controller (PLC) is becoming increasingly difficult. In the not too distant past, programmable controllers were limited to performing sequential control vis-a-vis turning on and off switches, motors, solenoids, lights, and instruments in preselected intervals or in combined sequence patterns controlling an operation (11,12). Recently, sequential controllers have been developed that can perform event-based control functions. These devices can be programmed to perform comparisons and combinational logic, as well as wait for specific events before proceeding to, or escaping from a point in the program if an event fails to occur.

Today, programmable controllers have moved far beyond their sequencing role, having evolved into rugged, versatile, powerful devices capable of performing sophisticated control tasks. Programmable controllers are now endowed with sophisticated arithmetic capability (and the associated scratch pad memory) and control algorithms (PID) to support continuous control applications, making them appear to be more like process controllers and can be used as such.

On the other hand, as was pointed out in the previous discussion of enhanced controllers, logic and sequential control capabilities are being added to these process control modules and, thus, they are taking on the characteristics of programmable controllers and can also be used as such.

So, where does the difference lie? Why has the effort been made to provide a means to interface the PLC to the distributed control system if it looks so much like the enhanced process controller. The answer, at least for the present, is application and the method of programming the PLC for this application.

PLC programming languages use functions related to machine and process control. Although various programming languages are being used today, relay ladder logic is by far the most prevalent language used. Unlike programming languages used in conventional computers, relay ladder logic is understood by most plant personnel and is easily learned even by those that have no programming training. Consequently, the use of PLCs in plants to replace outdated and worn out hardwire relay logic panels is very comfortable for the plant electricians and, therefore, a popular alternative.

If the PLCs are being installed as part of a plant modernization project which includes the installation of a distributed control system, then there is considerable advantage to being able to interface the PLCs to the communication network to take advantage of the operator console to display PLC data and diagnostics and the ability to download programs to the PLC from the host computer. The PLC program can be archived on the host's bulk memory device. Even if the PLCs were previously installed over a long period of time, the benefits from interfacing them to the distributed control system still are significant.

4.4 PROGRAMMABLE LOGIC CONTROLLER ARCHITECTURE

Essentially all programmable controllers, regardless of size, cost, or complexity, are designed around the same basic hardware elements and possess similar functional characteristics. The major components of a PLC are the central processing unit, memory, variable input and output processor and the power supply. Additionally, a PLC system will include various peripheral devices, such as programmer, operator's console, hard copy printer, a communications interface module and special function control modules. Figure 4.18 shows a layout of the major components for a programmable controller.

4.4.1 Controller

Most programmable controllers designs are based upon microprocessor chip technology and, thus, they function very similarly as the familiar microprocessor-based computer. The CPU is the logical device which organizes and initiates all controller activities. The CPU receives input data, performs logical decisions based upon a stored program, and generates a set of control actions in terms of values to be stored in the output driver.

The logical decisions performed by the CPU include straight logical processing, elapse time monitoring, event counting (contact closures on rotating machines, for example), output latching, comparing, and data storage. The elapse time required for the CPU to complete one operating cycle is defined as the scan time. The elapse time between the initiation of each scan is called the scan interval.

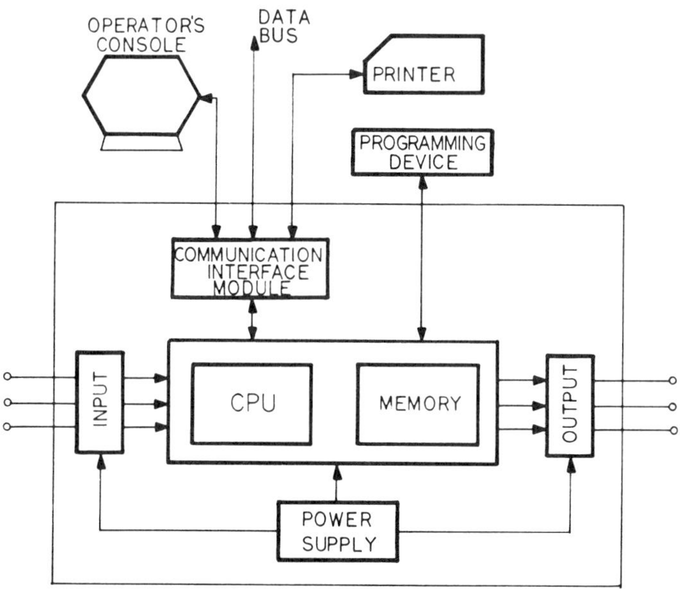

FIGURE 4.18 Functional architecture of a typical programmable controller system. The heart of the system is the central processing unit (CPU) with memory and signal input and output logic. The communication interface module supports communication to external devices.

During each scan time, which typically runs from 5 to 200 milliseconds, depending on the particular PLC model, memory size, and the complexity of the user logic being executed, the status of the input and output circuits will be scanned and the application program run, thereby determining the new values of the output circuits. The output circuits are then updated.

The PLC memory functions exactly like that for any microcomputer, that is, it provides a means for storing data, application program instructions, and the executive program. The executive program instructions, which direct the PLC's activities, are provided by the manufacturer. The user develops his desired control strategy, translates this into application program instructions, and loads these into the PLC memory.

The complexity of the control strategy, in terms of the number of input and output points, can be scanned and controlled and the sophistication of the logical relationship, determines the amount of application program memory consumed.

The function of the input module in the PLC is to convert high voltage signals from external devices to low voltage signals required by the CPU. Typical external field devices are push buttons, limit switch and relay contacts, analog sensors and pulse streams.

The CPU generates a set of output values after each scan cycle and transmits these values to the output relay modules, which in turn, develop signals that drive field devices such as motor starters, solenoid valves, indicator lights, positioning valves, and relay coils. The input and output modules usually consist of groups of input and output circuit cards mounted in a chassis. Typically, each card will handle 4, 8, or 16 input or output signals. The point capacity of a single PLC system will range from 32 to 4,096. Some PLCs will provide special cards to accommodate analog input signals provided by process sensors (thermocouples and transducers). These PLCs will also provide software algorithms to perform PID type continuous closed-loop control.

With recent trends toward installing distributed control systems in plants and the centralization of control, it has also become very desirable to be able to monitor the performance and status of PLCs already installed or being installed at a plant site. This can be accomplished in the same manner as the distributed controller, that is, install a simple communications interface module within each PLC and connect it to the plant data communications network. Operation of the PLC can then be monitored from a centrally located computer driven operator's console. Variant conditions will be alarmed and operator actions can be taken promptly. It even becomes feasible to downline load the PLC programs or to update existing programs or initiate changes in control strategies from the central control console once the PLC is on the network.

4.4.2 PLC Programmer

Cost effectiveness constraints have dictated, until very recently, that the PLC programming devices not be an integral part of the PLC, but a separate portable device. Programming units vary in size and capacity and, consequently, ease of use. They range in size from small portable hand-held devices about the size of a small desk top calculator to colored CRTs with graphics capability.

Programmers are usually only connected to the PLC while the program is being entered or updated. When this task is complete, the programmer is disconnected. The programmer, which has its own power cord and nonvolatile memory, permits the user to perform programming tasks in an office environment, and then load the completed program at the field site.

Recently, the trend has been to utilize medium size microcomputers with colored CRTs and sophisticated programming aids to program and document the PLC programs. These systems produce a finished program which has been configured, tested, and documented extensively. This reduces the probability for field installation problems and, thus, reduces field time and increases productivity of the user.

4.4.3 Peripheral Equipment

As programmable controller functionality has increased, it is migrating from the motor control center and backroom relay rack to the control room where it is now performing integrated logic and continuous control for many batch and semibatch type operations. With increased involvement in process control, it has become increasingly important that the operator be able to observe the performance of the PLC and not just its status. Consequently, the CRT terminal has become a significantly useful peripheral device for the PLC system.

The CRT terminal can provide visual displays of required operator actions and communicates general system operating conditions. In PLC systems, where continuous control functions are also implemented, the CRT is convenient for display loop status and developing trend displays.

Hard copy logs of PLC control activities may also be desired in the control room environment. This can be accomplished by adding a printer to the PLC via a communications module. Printers will range from full 80 to 132 column devices to relatively inexpensive strip or short-column printers. The requirements will dictate which capability is needed. If the printer is to be used in conjunction with programming and the generation of ladder diagrams, then the more expensive 80 or 132 column printer is advisable.

4.5 PLC PROGRAMMING TECHNIQUES

As was stated in the introduction to this section on programmable controllers, the key feature which distinguishes the PLC from other microcomputers is the very application-oriented programming languages available for the PLC user. The most popular and user friendly programming language for the PLC is relay ladder logic. This language is understood by most plant personnel and is easily learned without formal programming training, (12). Other languages used for programming PLCs are Boolean Algebra representations and BASIC.

4.5.1 Ladder Logic Programming

PLC ladder logic programming is based upon diagrams which are derived from series and parallel electromechanical networks. The ladder logic diagrams are structured, as were the electromechanical networks they represent, from combinations of series, parallel and combination series parallel circuits. Ladder logic shows the relationship between output actions as a function of input signals conditions and is drawn in a series of rungs. Each rung contains one or more inputs and the output or outputs which are directly controlled by the inputs.

Three basic logic functions, AND, OR, and NOT, are used to describe logic circuits no matter how simple or complex they are. Figure 4.19 shows a diagram of an electromechanical system for on/off control of a motor. The pushbutton B is normally open. When the pushbutton B is closed, the coil of relay R will be energized and the relay R will close the switch to start the motor. Figure 4.20 shows how this same logic can be represented using the relay ladder logic diagram utilizing the two basics for input and output. Typically all inputs to the circuit, whether internal or external, are represented by the symbol shown in Fig. 4.21. All outputs from the circuit, whether they be

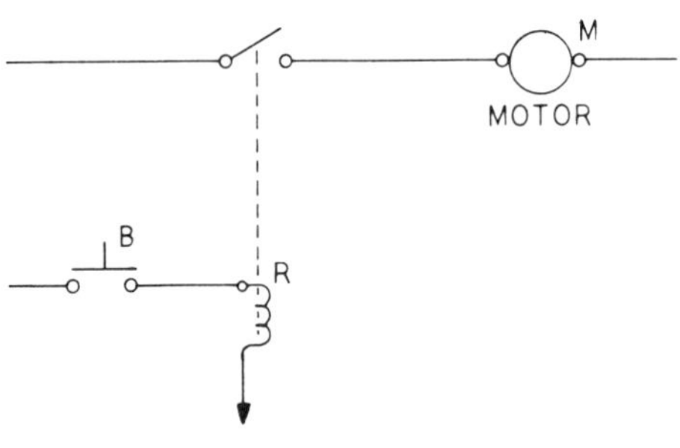

FIGURE 4.19 Electromechanical diagram for a motor start and stop showing the start button, B, the starting relay, R, and contact.

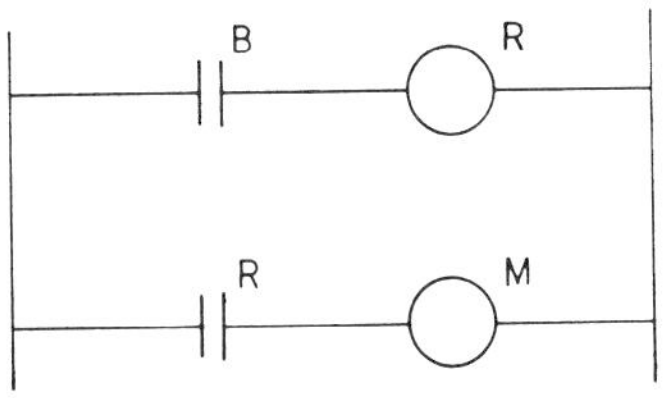

FIGURE 4.20 Ladder logic diagram of a motor start stop circuit showing symbols for start button, relay/contact, and motor.

internal or external, are represented by the symbol shown in Fig. 4.22.

4.5.1.1 AND Logic Configuration

An example of the use of the AND logic configuration is shown in Figure 4.23 in three different notations:

1. As an electromechanical diagram.
2. As a logic ladder diagram.
3. As a logic symbol. (Shown is the AND symbol typically used. No standard symbology has been agreed upon by the different user groups).

In this example the output state, L, is determined from the input states, A and B, according to the following truth table.

A	B	L
Open (Off)	Open (Off)	Off
Open (Off)	Close (On)	Off
Close (On)	Open (Off)	Off
Close (On)	Close (On)	On

A timing diagram of this circuit is shown in Fig. 4.24 to illustrate the dynamic effect of changes to the inputs on the output.

FIGURE 4.21 Ladder logic symbol for an input signal to the programmable controller.

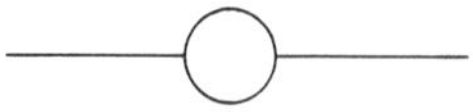

FIGURE 4.22 Ladder logic symbol for an output signal generated by the programmable controller.

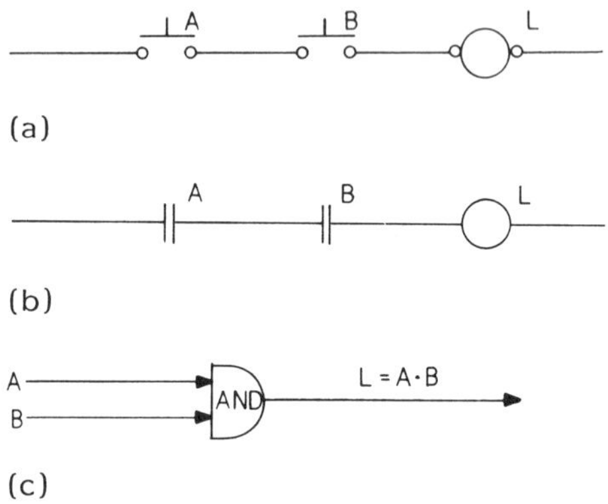

FIGURE 4.23 AND logic configuration shown in electromechanical, ladder logic, and Boolean algebra symbologies.

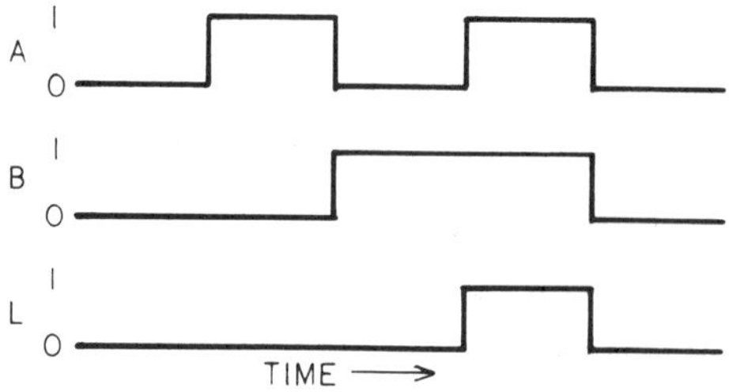

FIGURE 4.24 Timing diagram for AND logic showing logic states of output, L, for each configuration of logic states for inputs, A and B.

4.5.1.2 OR Logic Configuration

An example of the use of the OR logic configuration is shown in Fig. 4.25 in three different notations.

1. As an electromechanical diagram
2. As a logic ladder diagram
3. As a logic symbol

The truth table for the OR logic is shown as follows:

A	B	L
Open (Off)	Close (On)	On
Open (Off)	Open (Off)	Off
Close (On)	Close (On)	On
Close (On)	Open (Off)	On

A timing diagram for the OR circuit is shown in Fig. 4.26.

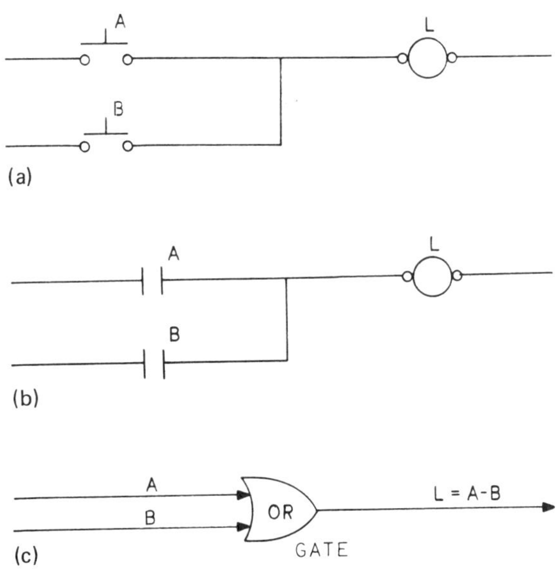

FIGURE 4.25 OR logic configuration shown in electromechanical, ladder logic, and Boolean algebra symbologies.

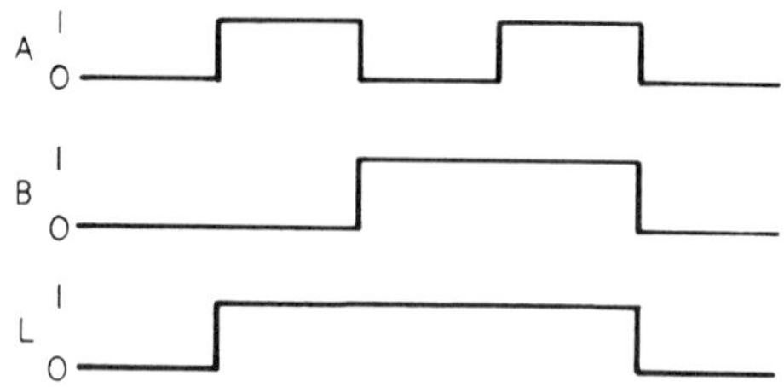

FIGURE 4.26 Timing diagram for OR logic showing logic states of output, L, for each combination of logic states for inputs, A and B.

4.5.1.3 NOT Logic Configuration

The NOT logic is used to activate the output of a circuit when the input(s) is turned off. Figure 4.27 shows an example of a circuit using NOT logic.

The NOT logic is usually used in conjunction with AND and OR logic to generate complex circuit designs. The truth table for the NOT logic is shown as follows:

A	L
Off	On
On	Off

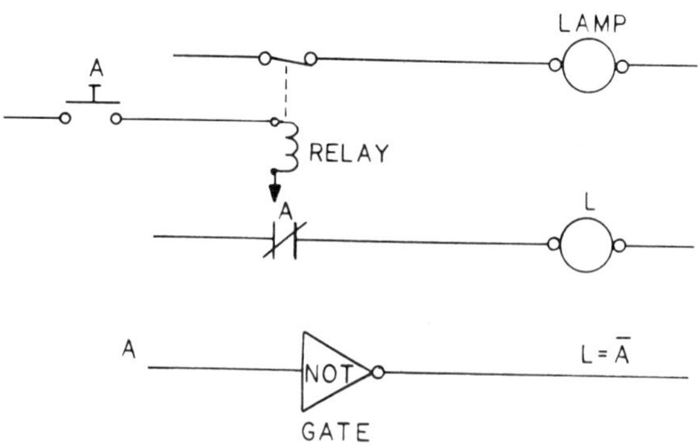

FIGURE 4.27 NOT logic configuration shown in electromechanical, ladder logic, and Boolean algebra symbologies.

The timing diagram for the NOT function is shown in Fig. 4.28. NOT statements are the logical equivalent of a complement statement in Boolean logic.

4.5.2 Boolean Algebra Programming

Boolean algebra is a shorthand notation for expressing logic functions. The Boolean equation for the AND, OR, and NOT logic functions are shown in Figs. 4.23, 4.25, and 4.27. Figure 4.28 shows a more complex circuit utilizing AND, OR and NOT logic.

The Boolean logic statement for this circuit is as follows:

$$((1PB \times 2CR) + 3LS) \times 4CR \times 5CR = SOLA \qquad (4.2)$$

For programmable logic controllers that accept algebra statements directly, the previous circuit would be programmed as follows:

LOAD	1PB
AND	2CR
OR	3LS
AND	4CR
NAND	5CR
STORE	SOLA

Consequently, the solenoid A can be turned on in either of two ways. (1) push button 1 and relay contact 2 are closed, plus relay contact 4 is closed and relay contact 5 is not opened; (2) limit switch 3 is closed, relay contact 4 is closed, and relay contact 5 is not open.

For those who are familiar with Boolean algebra, it provides a very neat and compact means for programming and documenting logic circuits.

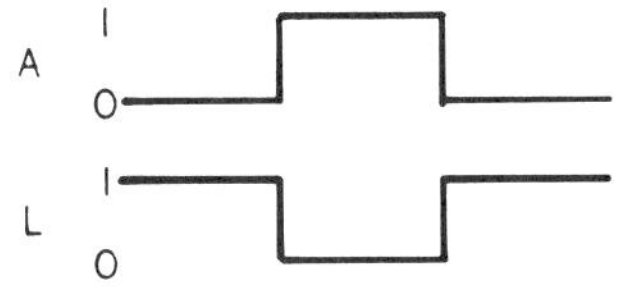

FIGURE 4.28 Timing diagram for NOT logic showing logic states for output, L, for each logic state of input, A.

4.5.3 BASIC Programming

Some manufacturers of more sophisticated programmable logic controllers are providing the BASIC programming language to enhance the ladder logic type of programming.

The addition of BASIC greatly enhances the functional capability of the PLC by adding arithemtic operations. This feature makes it possible to perform continuous three-mode digital control. It also permits interfacing with analytical instrumentation. The latter facilitates performance calculations and permits corrective actions to be initiated.

BASIC programming capability also makes generation of logs, alarm messages, and operator displays much simpler.

4.6 INTEGRATED CONTROL PHILOSOPHY

The first efforts to quantify the performance of an entire plant or operation became feasible in the early 1950s as a result of two coincidental events, the commercial development of the industrial digital computer and the evolution of the linear programming algorithm and the associated methodology for solving a complex set of linear inequalities. For the first time, it was possible to develop an economic model of the plant in terms of raw material costs and availabilities, activities, and product values and demands. Although severely limited to linear relationships, this technique provided a great leap forward in understanding internal plant economics and in making adjustments to plant operation based upon data acquired from quantitative analysis (13–15).

Beginning in the 1960s process control computers became an integral part of the plant control system interfacing with the analog controller and providing supervisory control based upon economic analysis of the plant operations in real time. However, even during this period, there were those who were talking the benefits of data logging systems. This approach provided an easy cop-out for those who had anxieties about going the whole way to closed-loop control. Just log all this data and some day, as if by magic, a revelation would provide new insight into the operation of the process. History has now documented that this approach most often proved to be an expensive mistake. It was a poor substitute for sound engineering analysis and good control practice.

As digital process control computers became more powerful and less expensive (in terms of capability), various approaches to developing integrated control evolved. In the late 1960s the General Electric Company developed large remote terminals which were located throughout the plant and connected locally to the process signals. These then communicated to a powerful central processor which actually performed the process analysis and control calculations.

Later the remote terminals were replaced by intelligent minicomputers which also communicated to a central supervisory computer. Functions performed frequently were processed in the minicomputer while the slower supervisory control functions, such as present-state calculations, economic evaluations, and optimization were performed in the central supervisory computer.

A parallel activity was the installation of large direct digital control computers which were directly interfaced with process measurement devices and sent signals directly to the control device. These systems were developed to provide control for several hundred loops. Even though the design concepts for direct digital control were sound, the implementation of the concept left much to be desired. The systems were very expensive and the control was unreliable due to the limitations in the computer design technology of that period.

Only with the introduction of distributed digital control technology has the implementation of a truly functional and effective integrated control philosophy been feasible. The earlier attempts to develop an hierarchical control system can now be fully implemented with definite clearly defined functions being performed at each control level.

In order to illustrate this concept let's consider a typical plant which is comprised of several operations (processes or units) which operate independently of each other, but which receive feedstocks from upstream (or parallel) processes and provide feed or products for downstream process or operations and, thus, are tied interactively with each other as far as overall plant performance is concerned.

For such a multiprocess plant an effective integrated control system should be configured as a three-tiered hierarchical system as shown in Fig. 4.29. Management objectives and goals for the operation of the plant enter the top tier. These objectives and goals can take various forms but typically are financial in nature. Typical goals would be: maximize returns on capital investments, maximize plant throughput, operate at production rate that is most efficient (minimize cost per unit production), or operate plant at planned rates based upon market conditions at minimum cost.

The objectives are then evaluated in the light of current plant performance data, raw material costs, product costs, and various constraints. The output of the first tier "optimization" (I hesitate to use this term since it implies a level of sophistication or complication which may not be warranted) is a set of plant operating conditions which are required to achieve the goals. These conditions are specific performance objectives for each operating unit in terms of what raw materials are to be charged, at what rates, to which units, and what product rates with what specifications are to be produced from which units.

The frequency of this calculation will depend completely on how frequently management or marketing goals change significantly. It may be performed daily but more likely weekly or monthly.

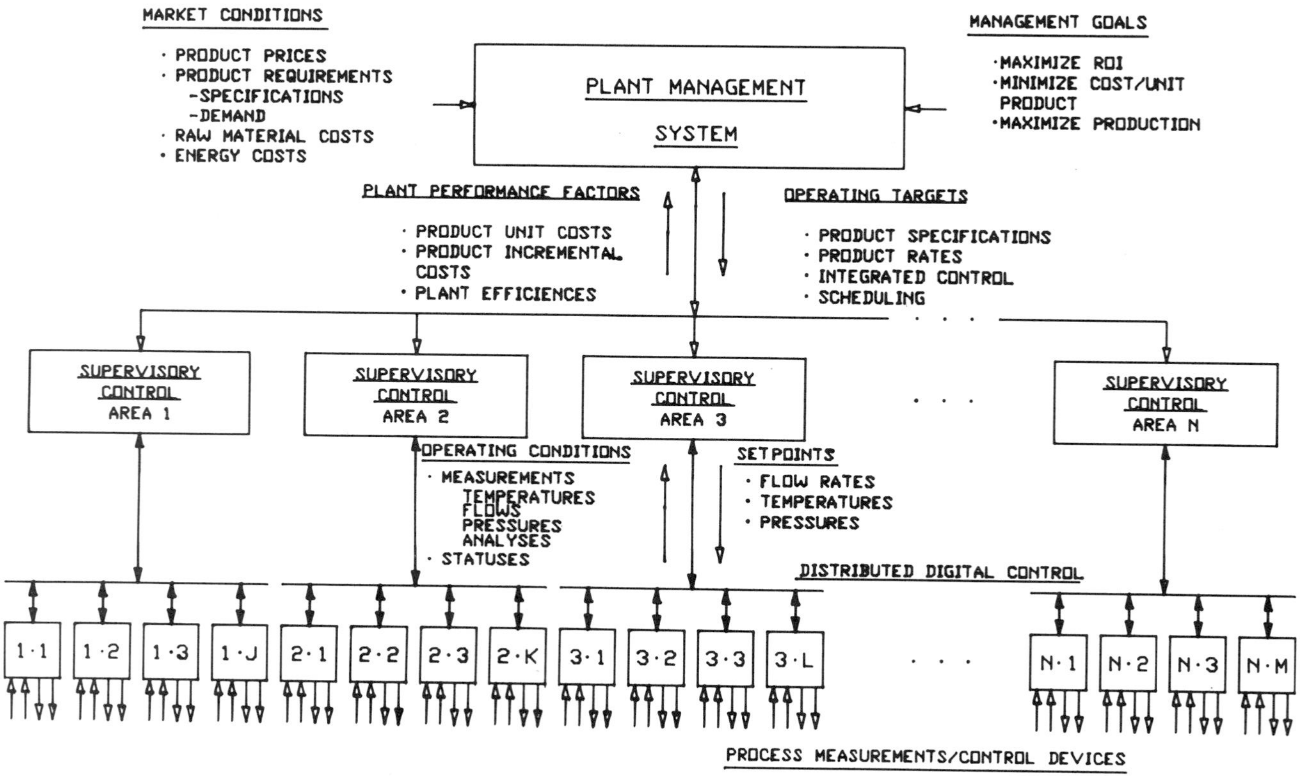

FIGURE 4.29 Block diagram of a multitiered plant management system showing information flow between tiers.

These performance objectives are then communicated to the second tier. The performance objectives for each unit are then translated using a process model into specific operating conditions which are required to meet the performance objective in the most efficient manner. This level of control is often implemented in the dedicated supervisory control computer associated with each processing area. The specific operation conditions are given such as: target feed rates, product rates, product specifications, level of conversion, reactor severity, flows, temperatures, pressures, etc.

The key point to be emphasized here is that this performance calculation is calculated at a frequency which is dependent upon the rate of performance variations within each process. When the performance change for a given process exceeds a certain threshold, then new target conditions should be calculated to again maximize the process performance. Changes in process performance are most often caused by variations in raw material characteristics but can also be caused by changes in catalyst activity, heater efficiency, heat exchanger efficiencies, etc. Therefore, this determination may be calculated as frequently as every ten minutes up to several hours depending on the frequency and magnitude of the change impacting the process performance.

The output of the second tier is in the form of setpoints for the key manipulatable variables for the process. In our opinion, these setpoints should be sent directly to the process control system which is the third and bottom tier in the control hierarchy. Experience has shown that is not practical to merely output operator guides and expect the operators to adjust the setpoints. After the initial implementation, the operator will soon ignore the computer generated guides, or be out to coffee, or even worse, try to second guess the computer, and input his own setpoints.

The bottom tier consists of the regulatory control which, until recently, was performed by analog controllers but is now being implemented in most new and many older plants with distributed digital control technology. The frequency of this control action is continuous for the analog controller based system and in the 0.5 to 30 second range for the digital control system.

The bottom level control system provides frequent control actions to respond to the effects of measured and sensed disturbances to the process operation and is designed taking process dynamics into account.

So to recap, management objectives and goals are translated to optimize operating targets in the top tier. These operating targets are converted into specific operating setpoints to achieve the operating targets for each process in the most efficient manner in the second tier. The setpoints are maintained via regulatory control to minimize the effects of disturbances on the process in the bottom tier.

Information about the current performance of each process in terms of yields, costs, efficiencies, is communicated up the hierarchy to provide a basis for the economic decisions.

REFERENCES

1. *Total Distributed Control TDC 2000—An Evolutionary Look at Process Control/1*, Honeywell Inc., Phoenix, Arizona, 1975.
2. *Total Distributed Control TDC 2000—An Evolutionary Look at Centralized Operations/2*, Honeywell Inc., Phoenix, Arizona, 1977.
3. *TOSDIC Distributed System*, Toshiba Corporation, Tokyo, Japan, December 1981.
4. *Taylor Introduces the MOD 30TM Control System*, Taylor Instruments/Combustion Engineering, Rochester, New York, November 1983.
5. *TDC 2000 Process Management Systems—Continuous Process Control*, Honeywell Inc., Phoenix, Arizona, February 1981.
6. *TDC 3000—The Totally Integrated Information and Control System*, Honeywell Inc., Phoenix, Arizona, 1984.
7. Higham, John, Kendall, Burton, Gerdts, Merrill, The Data freeway network: A coaxial multi-drop management control bus, *Control Engineering 28*(10):103–106, (September 1981).
8. *Total Instrumentation Control System One-Loop Controller TOSDIC-200 Series*, Toshiba Corp., Tokyo, Japan, September, 1981.
9. *Micon P-2000 Controller Function Description*, Company Bulletin p200-37-0, Houston, Texas, April, 1981.
10. *Bailey Network 90 Controller Module NCOM01/02*, Product Instruction E93-906, Bailey Controls Company, Wickliffe, Ohio, 1980.
11. Jannotta, Kenneth L., "What is a PC?", *Instruments and Control Systems*, February, 1980, pp. 21–25.
12. Meinhold, Ted F., Reference File: Programmable Controllers, *Plant Engineering*, November 23, 1983, pp. 51–67.
13. Lane, James W., Impact of digital control evolution on control system design philosophy, *Proceedings of the 1984 Canadian Conference on Industrial Computer Systems*, Ottawa, Canada, May, 1984.
14. Pujol, John B., Moving to millwide automation to ensure profit and survival, *Southern Pulp and Paper*, January, 1984, pp. 36–41.
15. Wage, Don, Production management systems: Where do we go from here?, *Proceedings of the 1984 Canadian Conference on Industrial Computer Systems*, Ottawa, Canada, May, 1984.

5

Process Control Computers

5.1 WHAT ARE PROCESS CONTROL COMPUTERS?

The broad definition of the term computer, according to Webster, is "one that computes." Thus, not all computers are electronic digital computers as they are usually defined today. Of course, the original computers were mechanical monstrosities designed around gears and cams. Then there has been the proliferation of the analog computer which is used to solve time related problems using resistors, amplifiers, etc. But some of the more interesting nonconventional computers are based on such exotic concepts as fluidics, optics, and chemical reactions. While it is not likely that any of these diverged paths will become the wave of the future, who can really say, especially reviewing the progress in integrated circuit technology over the past 20 years and the proliferation of the computer "chip."

For all practical purposes, digital computers are indeed dominating the market now and may become the only type of computer in the future. They are being applied everywhere these days, and it is safe to predict that more applications are coming in the future. Although all digital computers share the same fundamental technology, each type of application has different requirements. This chapter is written to describe the basic characteristics of digital computers, and how they are applied to the process control problem. Because this is such a broad subject, only important highlights, which may affect the decision-making for using a computer, will be discussed.

A digital computer is an electronic system consisting of logic gates which operate in a binary state. The so-called binary notation is the expression of a number in base 2 using only two digits: "0" and "1". The use of binary arithmetic in computers is convenient since the majority of electronic devices are designed to operate with two logic states: "off" (0) and "on" (1). By linking a number of electronic devices together, numbers can be stored and manipulated.

The decimal system expresses numbers to the base 10, giving number 0 to 9, and a multidigit decimal number is expressed in terms of 10 or its power. For example, the number, 543, can be shown as:

$$
\begin{aligned}
 & 5 \times 10^2 = 500 \\
+ & 4 \times 10^1 = 40 \\
+ & 3 \times 10^0 = 3
\end{aligned}
$$

Likewise, a multidigit binary number is expressed in terms of 2 or its power. For example, the binary number, 10110, can be shown as:

$$
\begin{aligned}
 & 1 \times 2^4 = 16 \\
+ & 0 \times 2^3 = 0 \\
+ & 1 \times 2^2 = 4 \\
+ & 1 \times 2^1 = 2 \\
+ & 0 \times 2^0 = 0 \\
\hline
 & \quad\quad\quad 22
\end{aligned}
$$

So the binary number, 10110, is just another form of expression for the decimal number, 22. Table 5.1 shows some corresponding binary numbers and decimal numbers. It is clear that when a sufficient number of digits are used, binary devices can be used to store or process any large numbers. However, binary numbers become very long and cumbersome for large numbers. Table 5.1 shows that the decimal number, 20, becomes a five-digit binary number, 10100. In order to simplify the use of binary numbers, the notation "octal" and "hexadecimal" numbering systems are used.

An octal representation of a binary number combines three binary bits to form one octal number, while hexadecimal combines four binary bits to one hexadecimal number. The relationship between octal, hexadecimal and decimal notations are also shown in Table 5.1. Note that letters A to F are used in hexadecimal because six more notations are needed on top of the 0-9 decimal notations.

In actual practice, all computers are designed by using fixed word length processors and memories. To understand this principle, a few simple definitions for terminologies are needed:

TABLE 5.1 Decimal, Binary, Octal, and Hexadecimal Numbers

Decimal	Binary	Octal	Hexadecimal
0	0	0	0
1	1	1	1
2	10	2	2
3	11	3	3
4	100	4	4
5	101	5	5
6	110	6	6
7	111	7	7
8	1000	10	8
9	1001	11	9
10	1010	12	A
11	1011	13	B
12	1100	14	C
13	1101	15	D
14	1110	16	E
15	1111	17	F
16	10000	20	10
17	10001	21	11
18	10010	22	12
19	10011	23	13
20	10100	24	14

Bit: A binary digit

Byte: Eight conjunctive bits

Word: The conjunctive bits which may be operated on by the computer at one time. (For process control computers it can be 1 to 4 bytes, or 8 to 32 bits).

K: One thousand. In many cases it actually implies 1,024 which is 2^{10}.

A 16-bit computer with a 128K-byte memory means the computer has a total memory of 131,072 bytes, and its arithmetic processor is 16 bits long.

A decimal number can be converted and stored on a computer in at least four forms: fixed point, floating point, ASCII (American Standard Code for Information Interchange), and BCD (Binary Coded Decimal). For process control applications, ASCII and BCD formats are used only for communications between devices, all calculations are made by using either the fixed point or floating point format. Detailed discussions about these formats will be given in later sections.

All process control computers have a special characteristic when compared to most other types of digital computers; they have the capability of processing external input and output (I/O) information in terms of digital or analog signals. That means a process control computer must have the capability of reading process measurement signals, such as the signal from a flow transmitter, and generating commands to control devices such as a controller or a valve. Because of this requirement, a process control computer is usually designed having hardware interrupt capability and can accommodate multitasking software systems.

5.2 COMPUTER HARDWARE DISCUSSION

A digital computer must have the following components: a central processing unit (CPU), clock, memory, input/output driver (I/O), and power supply. Figure 5.1 shows a simplified block diagram for a computer (1,2).

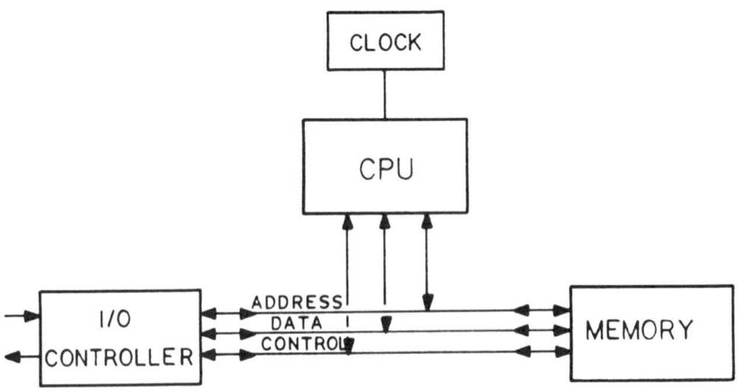

FIGURE 5.1 Simplified block diagram of digital computer with direct or memory mapped input/output with memory, central processing unit, and input/output controller sharing common bus.

5.2.1 Clocks

There are two types of clocks within a computer. The system clock is a high-frequency oscillator which paces the operation of the computer. The real-time clock is a timing device which tracks the time of day. Some computers may use the AC power as the real-time clock. This can be done by counting the number of pulses of the AC power and maintaining a "software" counter in computer memory. This clock requires manual initialization, that is, real time must be entered by the operator when the computer is powered up.

A faster system clock implies a faster computer when everything else is equal, but a fast system clock alone is not sufficient for fast computation speed. The clock cycle is the time duration required to complete one operation, and an instruction may require several clock cycles to complete, depending on the design of the CPU. Computer users need to be aware that a microprocessor with a five megahertz clock is much slower in computing speed than a large computer with the same clock rate, because large computers have more powerful CPUs in the logic sense.

5.2.2 Central Processing Units

The Central Processing Unit (CPU) is the "brain" of the computer. It consists of a program counter, an instruction register and decoder controller, an arithmetic-logic unit, accumulators, and registers. The CPU is designed to execute instructions sequentially at a very high speed. A modern CPU can usually process in excess of a million instructions per second. An instruction is not always an arithmetic operation. For example, "load the content of memory location X to the A-register" is an instruction. A simple calculation may require the execution of many instructions because of the software overhead and type of calculation.

One of the significant differences in CPU design is whether a separate "hardware" floating point processor is used. Detailed discussions will be made in the software section. A CPU with floating point hardware is very desirable when the unit is used for real number calculations, such as solving differential equations by numerical analysis. However, the floating-point capability will not help much when the unit is used for logic processing, such as required for a safety interlock systems.

5.2.3 Memory

There are two basic categories of memory used by a computer: main memory and bulk memory. Main memory are electronic devices which can be addressed by the CPU one word at a time. Bulk memory are those whose content must be transferred to the main memory and then be addressed by the CPU. Although most bulk memory devices are rotating devices with large capacities, this is not necessarily so. Although not a common practice, identical hardware devices have been used as main memory and bulk memory. The only differences is that those used as bulk memory cannot be directly addressed by the CPU.

5.2.3.1 Solid State Memories: RAM, ROM, PROM, and EEPROM

The majority of today's computers use RAM and ROM as their main memory. The term RAM (Random Access Memory) is usually limited to semiconductor read/write memory. The magnetic core types of memory, although it can be randomly accessed by the CPU, is not considered as RAM.

RAM has replaced core as the main memory for most computers because of lower prices and faster access times. RAM is made using semiconductor materials such as MOS (Metal Oxide Semiconductor). Today's integrated circuit technology already produces 256K-bit RAM memory chips in commercial quantity, and a one megabit chip is not far down the road. These large chips are being used for constructing large computers. For process control computers, 4- to 64K-bit chips are commonly used. Even with this "small" size, a so-called megabyte memory size may only required one to two boards in physical size.

A single RAM cell contains one or more transistors and one capacitor. When the capacitor is charged, it is at the "1" state. Since the charge on a capacitor leaks off with time, it requires power to maintain the "memory" state. When the power is cut off, the memory is lost. Some systems are designed to have a "battery backup" for the memory board. This battery will only keep the memory capacitors charged when the power supply to the computer is cut off, and should not be confused with the "battery back-up" system which has sufficient power to keep the entire computer system running during a power outage.

RAM can be read from and written to by the CPU. So this type of memory can be used to store programs temporarily or permanently. For example, programs can be loaded from a bulk memory to RAM at different times for different purposes.

There are programs which must be permanently stored in memory, and should not be changed or lost under any circumstances, such as the real-time monitor or language interpreter (both are to be described later). These programs are best stored in memory devices which cannot be altered. This type of memory is called ROM (Read Only Memory). ROM do not need a battery to retain their data when the power to the computer is cut off.

Sometimes a user wishes to store a program in permanent memory that cannot be changed for the purpose of protection. This can be accomplished by storing these programs on PROM (Programmable Read Only Memory). A popular form of PROM being used is the FEMOS circuit which is very similar to a RAM. This device can only be charged with a high voltage source. Thus, once programmed (using a special device), they cannot be changed by a normal computer "write," and the information will stay in memory for an indefinite time. This memory can be erased by exposing the chip to ultraviolet light for a period of time, thus, it can be reused.

Another type of PROM is the so-called EEPROM (Electronic Erasable Programmable Read Only Memory). By definition, it can be erased electronically. Both PROM and EEPROM are protected from power failure or erasure by accident, but they are also more difficult to change, thus, they should only be used for the storage of critical programs.

5.2.3.2 Bulk Memory

The most commonly used bulk memory devices for process control computers are magnetic discs. There are three types of discs, each appropriate for somewhat different purposes. These are: moving head discs, fixed head discs, and floppy discs.

A moving head disc is also called a hard disc. (This is to distinguish it from floppy disc which is made of flexible plastic film.) The moving head disc pack can be easily loaded into a disc drive and ready for operation in a few seconds. The relationship between a disc and a disc drive is just like that between a phonograph record to a record player.

A disc may have a storage capacity ranging from 10 to a 1,000 million bytes. Figure 5.2 shows a simple schematic for a moving head disc, which rotates at high speed, typically 3,600 rpm. Data are arranged in sectors and tracks, and the read/write heads do not touch the surface to read or write data on the disc. Since the heads have to be moved in and out to find the track, and must wait for the sector to rotate under the heads before data can be read or written, the elapse time between the CPU command to the actual read/write action can be fairly long, typically at about 25 milliseconds. This is called the disc access time.

A very large disc uses multiple layers of discs as one unit, and the read/write heads are moved in and out to read all layers. At each fixed location, the read/write heads can reach a large volume of data stored on a "cylinder" (identical track for all discs). It is desirable to keep all related information on the same cylinder to minimize access time.

A special type moving head disc called Winchester disc has been widely used with control computer systems. A Winchester disc may have 5 to 50 megabytes of memory and is usually quite economical. The major difference between a Winchester disc and other hard discs is that the read/write heads and the disc are sealed in a single unit, so the disc is not removable. This design improves the reliability of the device

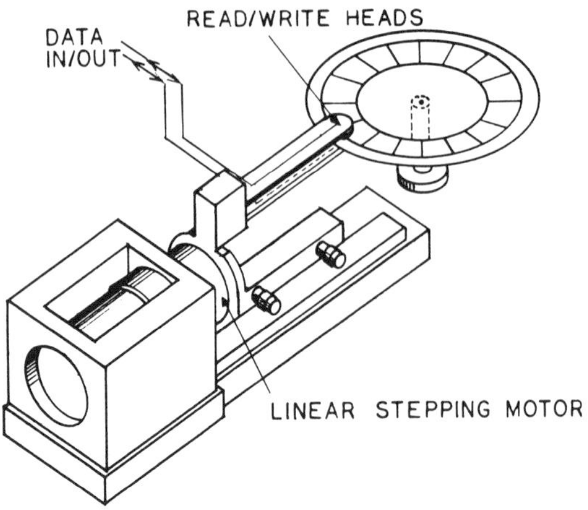

FIGURE 5.2 Moving head magnetic disk memory system. The simple system shown here has only one disc, but multidisc systems are common. Information is stored on both sides of a disc by separate read/write heads. A typical modern system has about 800 tracks per side, with storage for 100,000 bits per track. The total storage capacity of both sides of one disc is, therefore, 1,600 × 100,000 or 160 million bits. Sequential information can be transferred into and out of system at rates of 5 to 10 million bits per second. Average access time to a random location is 25 milliseconds.

because no dust can reach inside the Winchester disc, but reduces the flexibility because only one disc can be used per drive.

A fixed-head disc, by definition, is designed to have the read/write heads on a fixed location over a rotating disc or discs. Functionally, it works as if locating a pair of read/write heads over each track of a moving-head disc, so the access time is shorter, and the cost is higher. Like a Winchester disc, fixed-head discs cannot be easily removed, but since it has less moving parts, it is more reliable than any other types of discs.

Floppy discs are designed as low cost bulk memory devices for infrequent access application. A floppy disc resembles a paper-thin, small phonograph record encased in a cardboard enclosure. Each disc can store from 100,000 to over one megabyte, and works functionally as a moving head disc, but at a slower rate. The floppy disc rotates inside its cover at a speed of 360 rpm, and the read/write head contacts the surface of the disc. Floppy discs are not suitable for frequent access because of the wear factor.

Another type of disc which is now under development is based upon laser technology. At this time, only "write once" and, subsequently, "read-only" laser discs are commercially available. This disc will certainly have an impact on future computer applications because a laser disc can store many giga (billion) bytes of data, and it has a short access time. Should laser discs become reliable with read/write capability, then the concept of using bulk memory may have to be changed as such a large capacity becomes available.

5.2.4 Input/Output (I/O) Controller (3)

The function of an input/output subsystem is to coordinate and control the information flows between the computer and its external environment. As shown in Fig. 5.1, the I/O controller and memory share the same bus to communicate with the CPU. The bus is usually comprised of a large number of parallel paths to facilitate data transmission at high rates, and is usually built as a backplane board with other cards plugged into it.

From the hardware standpoint, bulk memory devices, such as discs, are part of the I/O subsystem. Bulk memory devices, as discussed in the previous section, are used for storing information. The other I/O devices are for facilitating external communication.

There are three basic types of I/O devices: peripherals, process inputs/outputs, and data communication devices.

5.2.4.1 Peripherals

The term *peripheral* has been used as a generic name for computer/human communication devices. These peripherals are used for loading programs and data into the computer or displaying results from the computer. The most frequently used peripheral device at this time is the Cathode Ray Tube (CRT) display with associated keyboard and printer.

Other peripheral devices, such as floppy discs and magnetic tapes, are also used for loading programs.

Like all electronic devices, peripherals change with technology. Not too long ago, punch card devices were popular peripherals, and now they are practically extinct because few people use punch cards any more.

New display devices, such as those using Liquid Crystal Displays (LCDs), could replace some CRT's, and other new devices, such as voice input, voice output, bar-chart readers, etc., could also be helpful for process control application.

5.2.4.2 Data Communication Links

Most modern digital control systems feature a multiplexed communication scheme which links computers and other digital devices by coaxial cables or a few pairs of wires. This type of serial communication links is called data highways, data path, or data way, etc., by various vendors. To implement a communications link, control logic must exist as part of the I/O drivers, and this hardware and software must be able to accommodate the necessary number of "talkers" and "listeners" at a reasonable rate of data transmission. The data transmission rate is measured by baud, which is defined as the number of bit signals transmitted in a second. A data highway with a one megabaud rate implies that this device can transfer a million bits, or roughly 130,000 bytes of information per second. This sounds like a large volume of data, but since there is always a significant overhead associated with data transmission to assure data validity, such as parity checks, high baud rate communication links are necessary to avoid delays, particularly when a large numbers of devices are connected to the link.

One important fact a user must be aware of is that high speed communication links obtained from different vendors are rarely, if ever, completely compatible. In other words, it is highly unlikely that one can directly connect a brand X computer to the data link of a brand Y computer without some interface hardware boxes and/or additional software. The cost and time requirements to implement this task could be much greater than most people expect.

Current trends in computer applications indicate that more communication links are going to be used in computer control systems because of the desire to implement plant-wide integrated information and control systems. To implement this type of communication network, it will become necessary to link several types of computers together. Some technical expertise in communications systems is needed to implement this task.

A new development in communication technology is the use of fiber optics instead of electrical cable. Digital electronic signals are converted to light pulses by a laser and transmitted to a receiver via glass fibers. This technology offers two major advantages—high data transmission rates and immunity to electrical interferences—as compared to existing technology. However, the application of fiber optics in process control systems has been slow because of the lack of low cost, reliable connectors and the high cost of the fiber optic systems.

5.2.4.3 Process Input/Output

The process computer is required to communicate with on-line instruments such as transmitters and controllers. Special I/O subsystems are required to handle the following four categories of process signals.

Analog inputs On-line process measurements generate three types of analog signals; current (typically 4–20 ma from transmitters), high level voltage (typically 1–5 volts from analyzers), and low level voltage (millivolts from thermocouples).

Analog signals must be converted to a digitized binary number before they can be read by the computer. Depending on the type of

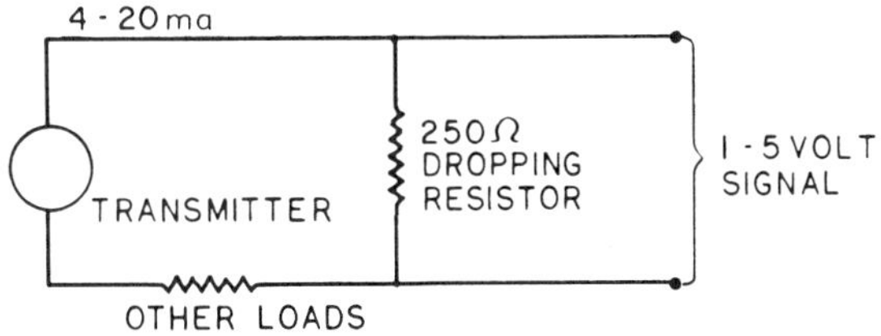

FIGURE 5.3 Schematic circuit for converting current signal generated by process transmitter into voltage signal for computer scanning.

analog input, different signal conditioning is needed prior to conversion. For example, a current signal will be converted to a voltage signal by passing it through a dropping resistor as shown in Fig. 5.3. The resistor could be part of an instrument or may be installed by the user. In either case, a high precision resistor is needed in order to produce an accurate reading.

Many analog signals have "common mode" noise, i.e., the voltage differential from each input wire to ground. The problem occurs in plants where high power electrical equipment are used. When this is the case, a low pass analog filter and an instrumentation amplifier, as shown in Fig. 5.4, should be used to reject the noise.

The analog low pass filter should not be confused with the digital low pass filter, which is to be discussed elsewhere. An analog filter is designed to reduce high frequency noise such as that resulting from 60 Hertz sources, while a digital filter is used to remove low frequency noise, typically, one cycle per second or slower.

Figure 5.5 shows an analog to digital converting system including a multiplexer, amplifier, and analog-to-digital (A/D) converter. The function of the multiplexer is the sampling of an individual analog signal by sequentially closing switches for a short period of time. Figure 5.5 shows a multiplexer with four input signal points. In real applications, the multiplexer can service from one to several hundred points, depending on the requirements for sampling frequencies and

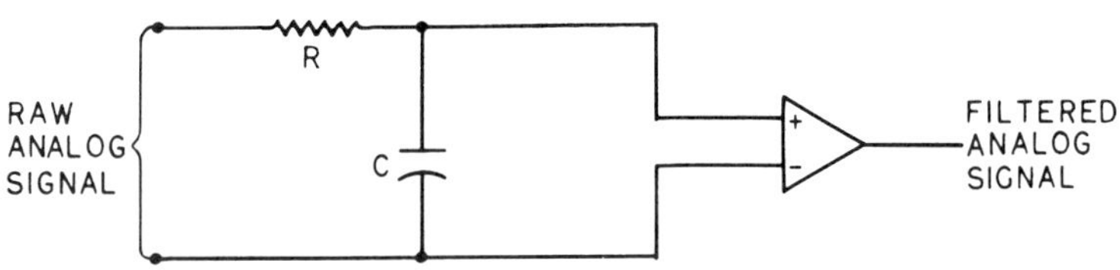

FIGURE 5.4 Schematic circuit of simple low pass analog filter to remove noise coexisting with raw analog signal producing a clean filtered analog signal for computer scanning.

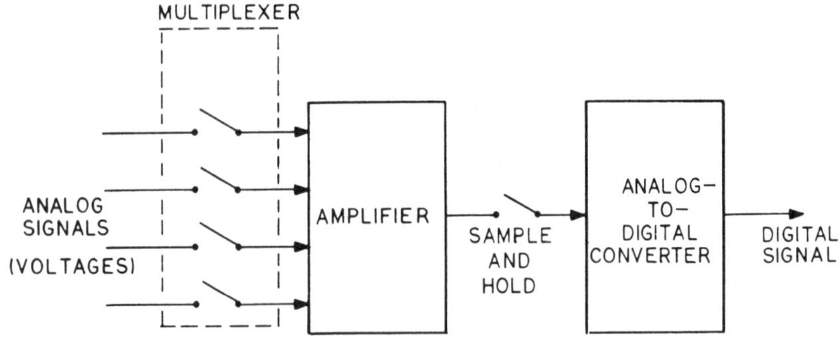

FIGURE 5.5 Analog input subsystem for processing multiple inputs through a shared amplifier and analog-to-digital signal converter.

desired reliability. The analog-input controller, driven by the computer, selects a point by closing the contact so that signal can be converted properly.

There are many different designs for analog-to-digital converters. The important consideration is how much resolution a particular A/D converter has. A 10-bit A/D converter plus sign bit has a range of 0 to 1,023 counts, thus, each count represents 0.1 percent of full scale. A 12-bit (plus sign) unit has a range of 0 to 4,095 counts, thus, each count represents 0.025 percent of full scale. Generally speaking, a 10-bit resolution is the minimum requirement and the 12-bit resolution or greater is desirable. Note that the 0.1 percent resolution is defined as 0.1 percent of the full value of the analog signal. The operating range of a measurement may be far below that. For example, a thermocouple may have a signal range of 8 to 12 mv, and it may be connected to an input amplifier which has a full input value of 20 mv input signal. The 10 mv signal will be converted to 512 counts. Suppose the 10 mv signal is 1,000°F, then each count of the digital signal will represent close to 2°F. This may not be sufficient resolution for operator display, and an A/D converter with a 12 to 14-bits resolution will be a more desirable choice. However, an A/D converter with greater resolution may not always be better. A 15-bit A/D may not be any better than a 14-bit unit because the last bit of the former device (indicating one part in 32,767) could represent just noise.

Digital inputs Digital inputs are those signals which have only two states, "on" or "off", such as the status switch of an electrical motor, or the "open" contact of a valve. A special type of the digital input is the pulse rate such as those generated by a turbine meter.

There are two ways the computer senses the digital input signals: by periodic polling of the status of the contacts, or by receipt of externally generated "change-of-state" interrupts. Periodic polling of the contact status is the most commonly used technique. The status of each point is scanned at a fixed time interval, typically, once per second. For the regular on/off status inputs, the purpose is usually to detect a change of state, so the computer just reads the points and saves the information for further processing. For a pulse input, such as the pulse stream from a turbine meter, the signal is usually connected to a hardware counter, and the computer reads and resets the counter, and accumulates the results. In addition, an overflow signal should be generated when the counter is full.

The interrupt is used to read the important externally generated change-of-state signals which require immediate attention, or when the time duration between status changes is critical. These inputs are connected to a hardware interrupt line, which interrupts the computer operation and requests that the inputs be processed immediately. Many sophisticated analyzers may be best served by connecting them to interrupt inputs.

Analog outputs Analog outputs from the computer are needed to generate signals for setpoint stations in the analog controllers, or to directly control valves, usually using current-to-pressure transducers, or to trend recorders. A digital-to-analog (D/A) converter is needed to translate a value in digital format to an analog signal.

In the early days of process control computer installations when the cost of digital hardware was expensive, a D/A converter was shared by many points by using a multiplexer. This multiplex concept is still being used, but many systems are using one D/A converter for each analog output in order to distribute the risk of a single D/A converter failure. A D/A converter can be designed on a single chip, and high cost is no longer a concern.

The resolution of a D/A converter, just like that for an A/D converter, depends on the number of bits used. For example, a 10-bit D/A converter has a resolution of 0.1 percent while a 12-bit D/A converter has a resolution of 0.025 percent. For closed-loop control, a 10-bit D/A converter should be sufficient because this works just like stepping a controller setpoint or a valve at 1,023 increments. This resolution has been proven to be sufficient for other types of systems.

The reliability of the D/A converter and the consequence of failures should be carefully analyzed by process control engineers before a system is selected. A D/A converter should be designed to allow users to specify the output value held upon the failure of the computer, such as 0 percent, 100 percent, or remain at last position.

Digital outputs Digital outputs are mainly used for controlling on-off devices, such as solenoid valves. Digital outputs are usually specified as either latching outputs, where the output signal is held after the computer command ceases, or momentary, where the output is released after the computer command ceases. Most output devices are

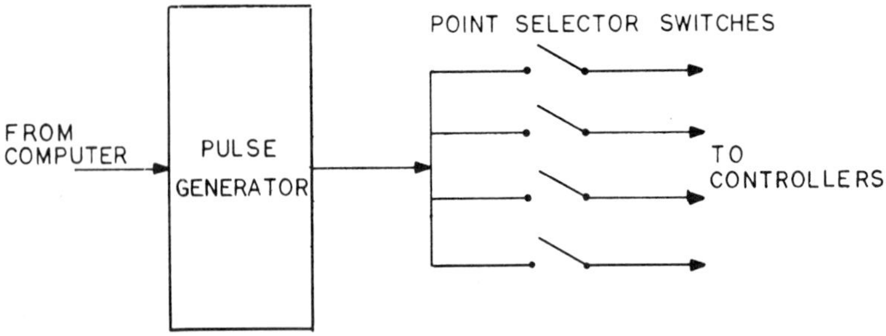

FIGURE 5.6 Commonly used method of converting computer generated setpoint value to output signal to analog controllers. Digital value is converted to equivalent pulse count.

designed utilizing solid state technologies such as transistors and thyristors.

A special type digital output device, the pulse output controller, is shown in Fig. 5.6. This device generates output pulses to four setpoint stations. Many older control systems were configured using this type of output device to drive setpoint stations equipped with stepping motors. This design has the particular advantage that the stepping motor could be installed in either a pneumatic or electronic controller and that the stepping motor did not move upon computer failure so the last setpoint value was always held. The operator could easily change the setpoint by jogging the stepping motor using instrument mounted push buttons.

Modern control systems are rarely designed using this type of output device because it requires a special software package. However, pulse output does offer one advantage over others. It will not change its position without power nor drift. Other electronic setpoint holding devices may change value upon failures of some components. When an output must be hold position for safety reasons, a pulse setpoint device should be considered.

5.3 COMPUTER SOFTWARE DISCUSSION

5.3.1 System Software

Without system software, the computer is a mindless electronic skelton with no purpose or direction. System software is the collection of basic programs that permits the computer to function as it is supposed. The executive program, input/output drivers, assemblers, compilers, etc., are all necessary parts of the system software.

The executive program, also called operating system, manages the computer hardware and other programs. It knows the size of main memory and how many and what peripherals are included in the system configuration, and it schedules which program is to be executed at a given time, depending on time and conditions.

Computer users should understand the fundamental differences between a single task, a foreground/background system, and a multiprogramming system. This will greatly reduce the chances for using a particular computer for the wrong job.

5.3.1.1 Single Task Executive

A single task software system utilizes the simplest executive and is capable of only processing one program at a time, and furthermore, this program cannot be interrupted by external events. Many personal computers on the market operate with this type of executive. Figure 5.7 shows a simplifed block diagram showing the allocation of memory locations for a single task executive program. (There could be many different ways.) In this case, all of the standard programs such as the "Bootstrap" load routine are stored in ROM. Users must load the desired program or programs from the disc to RAM by executing a load request via a keyboard. When the transfer is complete, the program is ready for execution. When the user desires to run another program, he must wait for the completion of the first program. This type of

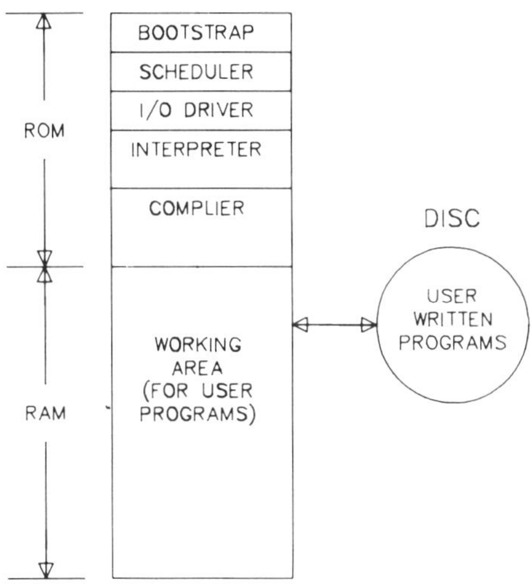

FIGURE 5.7 Memory allocation in single task executive program showing program assignments between ROM and RAM memory.

executive is simple and usually user friendly (impling that it is easy to use and understand), but it does not have much potential use for on-line, real time applications. The reason is that the single task system must perform each task in a fixed sequence without interruption.

5.3.1.2 Priority Interrupts

The term *priority interrupt* requires some explanation. A computer is a sequential machine. Once a program is started, the computer always executes the instruction stored at the very next location upon the completion of the first instruction (unless a branch instruction is used). An interrupt response software routine is designed to allow the computer to stop the execution of the program in progress, store the necessary intermediate information in memory, and initiate another program which has a higher priority and is assigned to that interrupt. After the computer has completed the higher priority program, it then retrieves the intermediate information, returns to the original program, and proceeds to complete that program.

Many process control applications require a computer with priority interrupt capability to effectively respond to changes in process conditions. For example, when a computer is used as a data acquisition system, which responds to the operator's request for displaying certain special displays, or performing certain special calculations, an executive system permitting interrupts should be used because the computer must promptly respond to the operator's requests.

5.3.1.3 Other Executives

A foreground/background executive is usually designed to handle priority interrupt routines. Figure 5.8 shows the memory allocation for a foreground/background executive program. (Again, this is just one of the many ways to allocate memory.) Some foreground programs that must be executed frequently can be stored permanently in RAM. The rest of the programs, both foreground and background, are stored in disc. Requests for transferring programs from disc to RAM are managed by the executive program, and these requests can be initiated by interrupt response routines or by the operator's request. Only one program at a time can be transferred to the work area. The highest priority program will be processed first. When it is completed, the next level program will be processed. When the computer responds to an interrupt, the programs will be transferred into and out of the memory according to the priority level. A common area can be reserved in main memory to store frequently used data.

The so-called background program is the lowest priority program and is usually assigned to a restricted area in RAM. This program is processed only when no other program is active. A system with back-

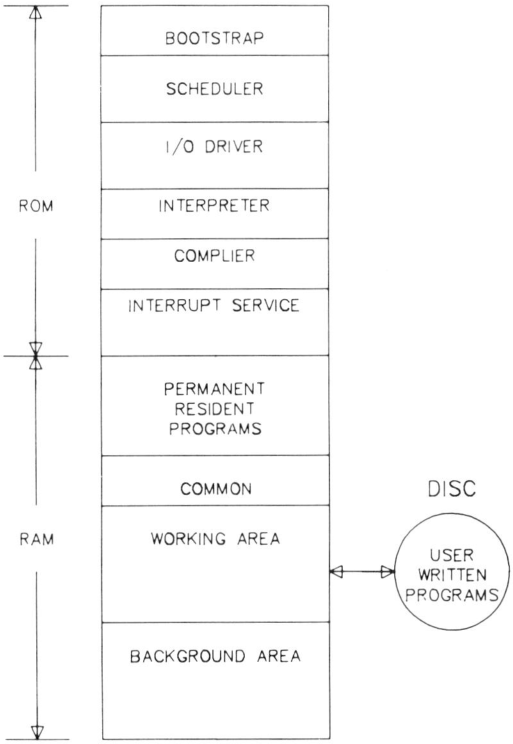

FIGURE 5.8 Memory allocation for foreground/background executive program showing program assignments between ROM and RAM memory.

ground processing capability is convenient to use when engineers wish to test a program using on-line data.

The multiprogramming executive program is the most sophisticated. This type of system allows multiple programs to be loaded into the working area, and maintains them there until other high priority programs need to be transferred into the memory. The multiprogramming executive is more efficient because, when properly implemented, the frequency of program transfers from disc to main memory can be reduced, and the computer can respond to outside events more quickly.

Executive programs and other system software are developed and provided by the computer manufacturer. Computer users should learn how to select a system to fit their particular requirements, but should not attempt to develop or alter standard software. System software may take years of operation to be fully debugged.

5.3.2 Data Handling

5.3.2.1 Arithmetic Calculations

A computer performs arithmetic calculations according to either of two formats, fixed point and floating point. For a given calculation, using the fixed point format is generally faster and more efficient in regards to memory usage than when using the floating point format, but the floating point format is more universal and less likely to cause problems for the user in the long run.

To aid in understanding how a calculation in the fixed point format is performed, the following example is given using a 16-bit word length computer. One single memory "word" can contain 65,535 "counts". These counts can be used to represent integer numbers ranging from −32768 to +32767. When more digits are needed, a double word or triple word format can be used. A double word (32 bits) format has a range of over ±2 billion counts while a triple word (48 bits) format has a range of over ±140 trillion (10^{12}) counts.

The triple word fixed point format is rarely used for process control applications. In fact, many control system suppliers use a single word fixed point format for both integer and real number calculations. This type of calculation has the advantage of faster execution speed and less memory consumption, as compared to the floating point calculations to be discussed later. However, fixed point calculations for real numbers requires the scaling of all numbers. To accomplish this, the range of the answers must be known before the calculations can be started. Consequently, for process control engineers, it is strongly recommended that a floating point format be used for all real number calculations, because the savings in memory using fixed point arithmetic cannot justify the cost of the increased manpower required to scale these numbers.

To appreciate the work required to perform a fixed point calculation for real numbers, consider the following simple calculation.

$$y = a\,x + b \tag{5.1}$$

A binary point or scale factor must be selected for each variable or constant. This binary point is the location of the decimal point in the 16-bit binary number. Assume the absolute value of y never exceeds 100, then the scale factor for this number can be selected by dividing 32,768 by 128 which is the next power of 2 greater than 100. Thus, 32,768 divided by 128 equals 256 (the number 256 equals 2 to the 8th power. It is convenient to use the powers of 2 as scale factors). That is, the number of counts, y_c, is equal to the actual value of y times 256. Assuming the value of y is 98.56, that is

$$y_{value} = 98.56 \tag{5.2}$$

then in fixed point format

$$y_{count} = 98,56 \times 256 = 25,231 \text{ counts} \tag{5.3}$$

Likewise, given:

$|a| < 4$

$|x| < 20$

$|b| < 20$

Then the scale factors for these numbers can be chosen as 8,192, 1,024, and 1,024, respectively.

With all variables and constants expressed in counts, Eq. (5.1) can be shown as:

$$y_c = \frac{1}{32,768} a_c \times x_c + \frac{1}{4} \times b_c \tag{5.4}$$

The constants (1/32,768) and (1/4) are the converting factors required to equalize the scale factors of the calculation, and these factors must be determined by whoever writes the program.

To check the calculation, assume

$a = 2.65 \quad a_c = 21,708$ counts

$x = 15.60 \quad x_c = 15,974$ counts

$b = 12.50 \quad b_c = 12,800$ counts

$$y_c = \frac{1}{32,768} \times 21,708 \times 15,974 + \frac{1}{4} 12,800$$

$$= 10,582 + 3,200$$

$$= 13,782 \text{ counts}$$

or

$$y = \frac{13,782}{256} = 53.83$$

Using a decimal calculation:

$$y = 2.65 \times 15.6 + 12.5 = 53.84 \qquad (5.5)$$

The difference in the calculated results is caused by the truncation of insignificant bits in the fixed point calculation.

Note that when the ranges of the variables or constants are changed, the scale factors must be changed. Otherwise, there will be overflow in calculation or insufficient accuracy.

For real number calculations, a floating point format should be used. There are many different type of floating point formats used by different computer vendors. Figure 5.9 shows a widely used format which uses two 16-bit words to show one number. The mantissa is 24-bits long, so the resolution is seven digits and the range is between 10^{-79} to 10^{75}. A double precision floating point format using three words or four words has a much higher resolution and range, but the two-word format should be sufficient for most process control applications.

Some computers use a three-byte floating point format, which uses 16 bits to show the mantissa and seven bits for the exponent. While this format covers almost the same range as the two-word format does, it has only a four-digit resolution. This may not be sufficient for some sophisticated calculations.

Calculations using a floating point format do not require scaling. The factors one should consider when using this format are range, resolution, and probably computation speed.

Many low cost computers are not designed with an integrated hardware floating point arithmetic unit. Thus, floating point calculations are made by using software subroutines, which can be 50 to 100 times slower than equivalent fixed point calculations. For an application which requires a large number of calculations, such as a frequent matrix manipulation, a hardware floating point coprocessor should be used. In fact, a floating hardware processor with current technology is not a costly item and should be considered as standard equipment for most systems.

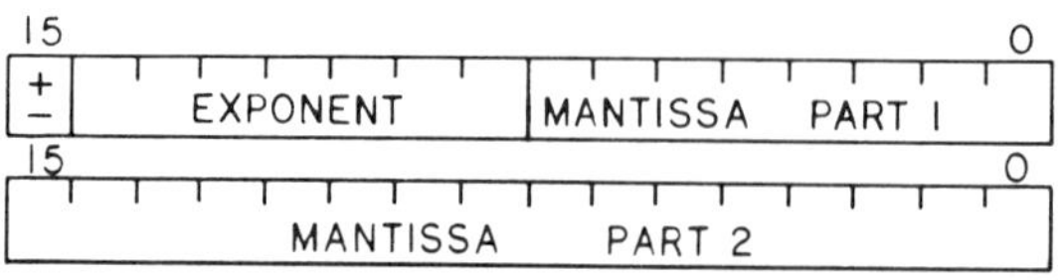

FIGURE 5.9 Floating point number format using two 16-bit words for sign, exponent, and two-part mantissa.

5.3.2.2 Data Communications

When the results of a calculation are transferred to another computer, they can be transmitted in binary form or as code characters, such as ASCII (America National Standard Code for Information Interchange). The binary form is more efficient because a number can be transferred using the least number of bits, so a link with a given baud rate can transfer more information per unit of time. But since the information that needs to be transferred can be a mixture of numbers, letters, and other characters, and many different types of fixed point and floating point formats are used, some identifiers and coder/decoder characters are needed to facilitate efficient data transfer. The development effort for interfacing two computers only using binary communication can be very difficult and time consuming.

The use of codes, such as ASCII, for communication can be simple when the *same* ASCII codes are used by sender and receiver. Table 5.2 shows two types of ASCII codes. Both are shown in hexadecimal numbers (the eight-bit format is used to check parity). To add to the complexity, some systems use a trimmed six-bit code to conserve memory. Therefore, even two systems, both using standard codes, may require an interpreter to communicate to each other.

Communications between computers is becoming more important, because when multiple computers are used, they require information from each other to be effective. The ability to communicate should be a major consideration when purchasing a new system.

5.3.3 Programming Languages (4)

A digital computer can only execute machine language code, that is, a code in binary number form to tell the computer how to perform its functions, and all data must also be in a binary form. In fact, early computer programmers did use machine languages to develop programs, and it was indeed a difficult task to write a program.

To ease programming efforts, computer manufacturers have developed different languages. These languages are special programs which convert different types of codes to machine languages.

There are two major categories of programming languages. The so-called low-level category includes machine languages and assembler languages which are similar to machine languages, but use symbolic representations for codes and labels for branch locations. The so-called high level languages are those using "English-like" statements and mathematical equations for programming. High-level languages are generally easier to learn, require less manpower to complete a task, but require more memory space and greater time for executing a program.

Computer users rarely have the need to use low-level languages for developing new programs, but may need to have a reasonable level

TABLE 5.2 ASCII Conversion Table

Graphic	8-bit ASCII code	7-bit ASCII code	Graphic	8-bit ASCII code	7-bit ASCII code
SPACE	A0	20	@	C0	40
!	A1	21	A	C1	41
"	A2	22	B	C2	42
#	A3	23	C	C3	43
$	A4	24	D	C4	44
%	A5	25	E	C5	45
&	A6	26	F	C6	46
'	A7	27	G	C7	47
(	A8	28	H	C8	48
)	A9	29	I	C9	49
*	AA	2A	J	CA	4A
+	AB	2B	K	CB	4B
,	AC	2C	L	CC	4C
–	AD	2D	M	CD	4D
.	AE	2E	N	CE	4E
/	AF	2F	O	CF	4F
0	B0	30	P	D0	50
1	B1	31	Q	D1	51
2	B2	32	R	D2	52
3	B3	33	S	D3	53
4	B4	34	T	D4	54
5	B5	35	U	D5	55
6	B6	36	V	D6	56
7	B7	37	W	D7	57
8	B8	38	X	D8	58
9	B9	39	Y	D9	59
:	BA	3A	Z	DA	5A

TABLE 5.2 (Continued)

Graphic	8-bit ASCII code	7-bit ASCII code	Graphic	8-bit ASCII code	7-bit ASCII code
;	BB	3B	[	DB	5B
<	BC	3C	\	DC	5C
=	BD	3D	]	DD	5D
>	BE	3E	↑	DE	5E
?	BF	3F	←	DF	5F

of understanding about low-level languages for troubleshooting of existing programs, and for defining specifications that a professional programmer can use to develop new programs.

High-level languages also have the advantage of being more machine independent, that is, a language used for one machine can often be used for another machine with little or no modification. These are also more convenient to maintain and usually provide more comprehensive documentation.

There are two types of high-level languages: compiled and interpretive. A program using a compiled language, such as FORTRAN, must be translated to machine code using a "compiler" before it can be executed. Both the source codes and the binary codes are usually stored in bulk memory, but the source will not be used again unless the program requires modification. Any modification of the source requires the recompilation of the entire program. When using an interpretive language the source codes are stored in memory and the program is executed one line at a time by converting each line to binary code and then executing the code.

A program written in an interpretive language can be modified and tested without recompiling. When developing new programs, this would be a major advantage. However, once developed, a program in interpretive language requires longer time to run.

Another form of high-level language is the so-called. "fill-in-the-blanks" software packages designed for process control (5,6). This type of system is much more complicated than other types of high-level languages, and the application of these is much more restrictive. The fill-in-the-blank systems are generally designed to allow a user, without programming expertise, to specify key software control parameters and configure the input/output variables. He can also determine how these variables will be operated on (selecting preprogrammed control algorithms), and develop operator communication in the form of CRT display and printed logs.

The fill-in-the-blank type of software system requires more computer resources (in terms of hardware and software) as compared to implementing the same functions by developing new dedicated programs, but also requires much less programming manpower and can be implemented within much less time. In addition, fill-in-the-blanks software systems are more engineering oriented, thus, they can be implemented by process control engineers with some basic knowledge of computers, and they can be maintained by engineers.

5.3.4 Developing a Computer Program

When developing an on-line program for performing a process control function, one is required to learn a computer programming language. However, knowing a computer language alone is not nearly sufficient to develop a program. One must also have a clear understanding of the problem to be solved, and the computer system software to be used. Because of the unique nature of real-time processing and operator communication, many tasks must be performed before any code can be written. In fact, these tasks are more time consuming and require more technical expertise than the coding of programs. These important tasks are process analysis, broad system specification, and detailed system design. The following example can be used to demonstrate these tasks. Note that this problem is unrealistically simple and is only used for demonstration of the procedure. Consider programming a small computer to control a mass flow loop as shown in Fig. 5.10. The control system should perform limit checks and warn operators via a CRT. The operator will use the CRT keyboard for data entry.

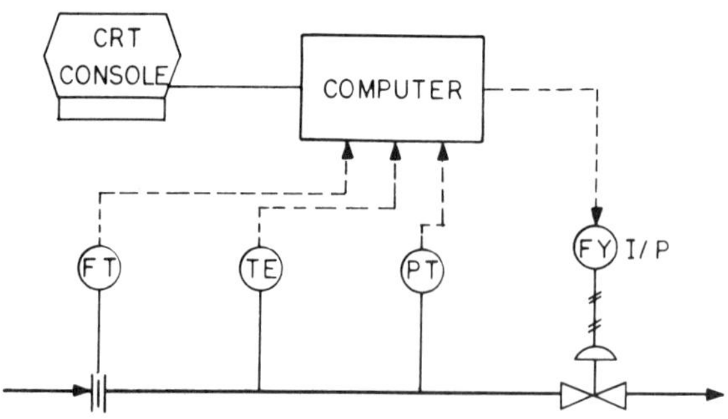

FIGURE 5.10 Computer system interfacing with process instrumentation required to control one loop.

TABLE 5.3 Tasks and Functions for Figure 5.10.

Task 0: Initialize the System

Task 1: Main Line:

1.1 Scan: Read analog input

1.2 Data conversion and limit check: (DATACON)

Check inputs for reasonableness
Filter inputs
Convert input to engineering units
Check process limits and set alarm

1.3 Control calculation (CALC):

Compute controller output

1.4 Output:

Check output for reasonableness
Send output

Task 2: Alarms: Sound and display alarm and respond to operator acknowledgement

Task 3: Operator Interface

3.1 Display

3.2 Data input

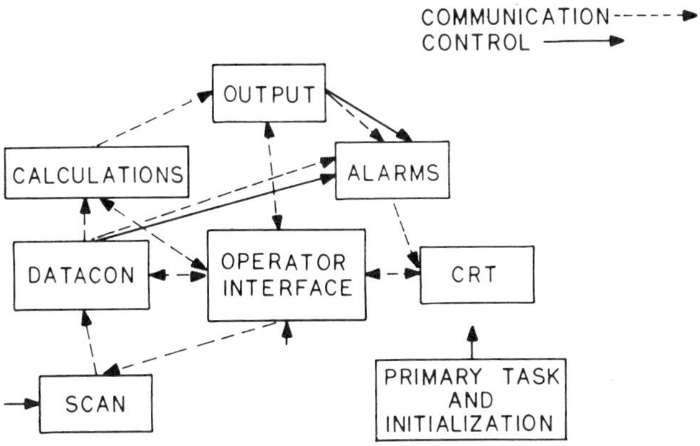

FIGURE 5.11 Program intercommunications for single control loop computer systems.

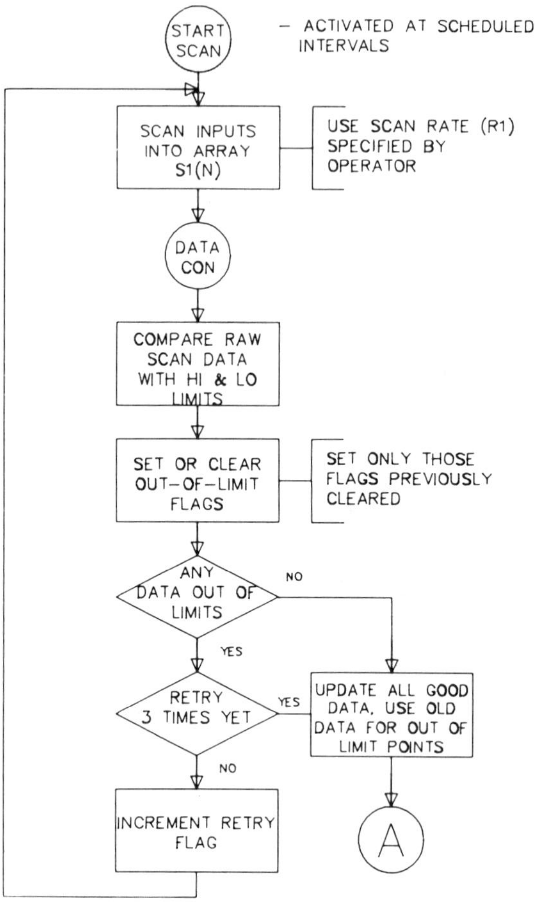

(a)

FIGURE 5.12 (a), (b), and (c): Mainline flow charts showing the procedures for scanning, filtering, limit checking of process variables, and for computing the controller output.

To start this program, the engineer must define the type of instruments to be read by the computer and the alarm limits. He then must determine the sampling frequencies, filtering constants, and specify the computation and control algorithms. The range of the output to the process must then be specified. The CRT display formats and the data entry procedures also need to be defined.

The second step is to develop a broad system specification to perform the defined tasks. This step identifies how many tasks are to be

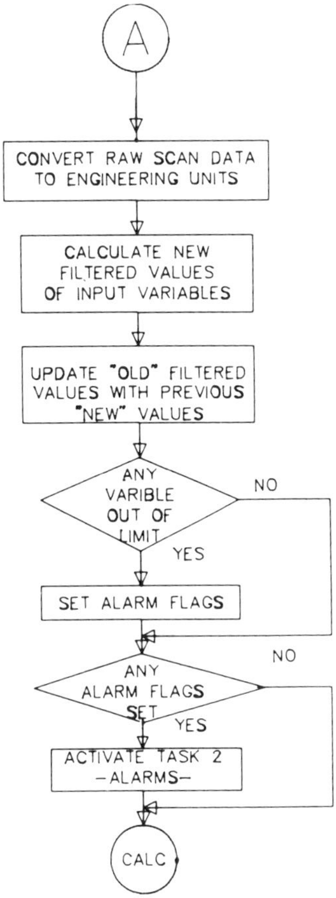

(b)

FIGURE 5.12 (Continued)

programmed, the priority level of each task, and how these tasks are to be linked as one system. Table 5.3 shows the tasks (or programs) and their functions, and Fig. 5.11 shows how these programs are linked together. Following the broad specification step is the detailed system design, which includes the identification of all address, parameters, flags, data file formats, and finally, flow charts.

Figure 5.12 shows the flow charts for part of the MAIN LINE program. This MAIN LINE program is only a small fraction of the overall system shown in Fig. 5.11. The largest task is the operator interface task, which requires many pages of flow charts.

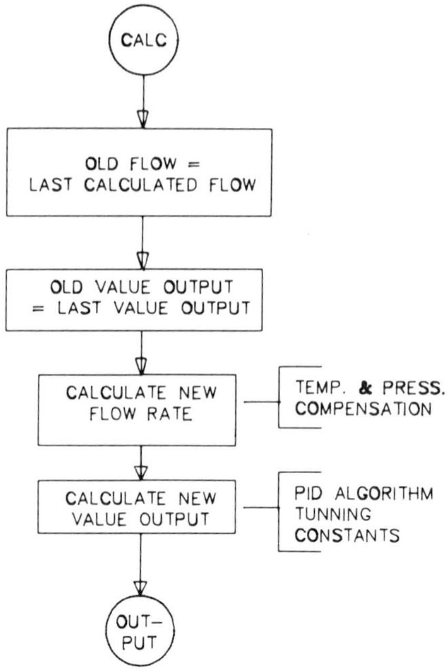

(c)

FIGURE 5.12 (Continued)

All of the programming tasks discussed so far are language independent, that is, the flow charts can be converted to codes using just about any language the computer supports.

The purpose for the previous example is to illustrate that the programming of a real time system requires far more knowledge than just an understanding of programming languages. This is the main reason that the cost for developing new computer programs usually exceeds the estimation made by managers.

Microcomputer-based digital controllers are designed so they may be configured by an engineer without having a knowledge of programming. The example described above can be implemented by just configuring a controller in the manner described in the previous chapter.

Many fill-in-the-blank software systems are designed to reduce the broad system specification efforts and to totally eliminate detailed system design, coding, and debugging. Users still must learn how to use these system, but most of them are not difficult to learn.

One shortcoming of the fill-in-the-blank systems is that no system can be so flexible that it can be used to implement every desired function. Sometimes, it is still necessary to develop new programs which have to be interfaced with the fill-in-the-blank system used. This may require some extra effort.

Regardless of the types of systems or languages used, programming is more analysis than coding. As the progress of computer technology continues, more advanced languages will become available, and coding may become even less of a problem, but analysis of a problem will always require special expertise.

REFERENCES

1. Korn, G. A., *Minicomputers for Engineers and Scientists*, McGraw-Hill, New York, 1973.
2. Bishop, R., *Basic Microprocessors and the 6800*, Hayden, New Jersey, 1979.
3. *PDP-11 Microcomputer Interfaces Handbook*. Digital Equipment Corporation, Massachusetts, 1983.
4. Ogdin, C. A., The highs and lows of microcomputer programming languages, *Instrument & Control Systems, 51*(5):65–68 and *51*(6): 81–84 (May-June 1978).
5. Tsai, T. H., "Software for process control computers," *Proceedings of the 1976 Computer Interface Instrumentation Symposium*, Instrument Society of America, Newark, 1976.
6. Alexander, J., Software packages for process control, *Advances in Instrumentation*, Vol. 39, Instrument Society of America, North Carolina, 1984.

PART II

Process Control Applications

6

Process Modeling

6.1 INTRODUCTION

Modern manufacturing generally can be classified into a number of major industrial sectors such as oil, chemical, power, gas, paper, cement, textile, plastic, auto, steel, machine, aerospace, electronics, and computers, etc. The common characteristic of these industries is that they have production plants. Each plant is an aggregate of the controlled manufacturing processes which are interconnected to form a production facility. To improve the productivity of the existing plant, there are basically two different approaches from the process plant point of view.

One is to modify or redesign the process in a more efficient manner. This approach can lead to either a different operational procedure, or a change in equipment utilized. Therefore, the changes to the process become more fundamental, and the investment can be more substantial. The second approach is to modify the existing operational procedure or control strategy without changing the existing plant. Therefore, the physical changes to the process or equipment become minimal, and the investment can be relatively small.

Both approaches require some type of mathematical representation or model as a design basis to generate the necessary information for improvement to the process. Because the emphasis in each approach differs, the mathematical models of the process, utilized for achieving such objectives, tend to be formulated in quite different manners (1-3). For process design, the steady state model is practical to use; however, for process control design the process dynamic model is more practical than the steady state model.

If the mathematical model derived from the physical or chemical properties for a given process is reasonably close to the reality, then the information produced from the model will be close enough for approximation of the actual process in operation. The advantage of the mathematical model is its capability to produce approximate information so quickly and inexpensively without operating the actual plant to obtain it.

For studies evaluating the benefits to be obtained from computer control of existing chemical plants, the emphasis is directed more toward the second approach. That is, the improvement of productivity results directly from control strategy or operational procedure changes executed by the computer. Therefore, in this chapter, the modeling techniques and model constructions which are utilized to support the derivation and evaluation of control strategy or operational procedure changes to a process are emphasized.

When an engineer is assigned a project for improving the operating efficiency or productivity of a process plant, his immediate reaction may be that he is overwhelmed by the complexity of the problem involved. The typical state equation for process, ($x = f(x, u, t)$), popular in the control engineering text books (4–6), is not available for most of the plant operations. Therefore, the direct application of optimal control techniques is not straightforward. Consequently, a practical method for constructing a meaningful model to facilitate control strategy testing and evaluation becomes an important subject.

6.2 MODEL USED FOR PROCESS CONTROL

Models utilized for operational improvements through the use of feedback control procedures emphasize several aspects, such as time or rate of change, which differentiate this type of model from the steady state models used for process design or optimal equipment sizing (7–9).

6.2.1 Input/Output Relationship

In general, the mathematical model used for control purposes should identify the process variables (such as temperature, pressure, flow rate, composition, etc.) to determine which of these variables are input to the process or output from the process (10). This is shown in Fig. 6.1. Input as defined here are the variables which can influence the process response; output is defined as the outcome of the process response. For example, the simple distillation column, shown in Fig. 6.2 is considered here to illustrate the meaning of the input and output (11).

The feed flow F, the feed temperature T_f, the feed composition $x_{fi}(i = 1, 2, . . . n)$, reflux flow F_r, heat input at reboiler Q_r, and heat removal at overhead condenser Q_c, can be considered as input to

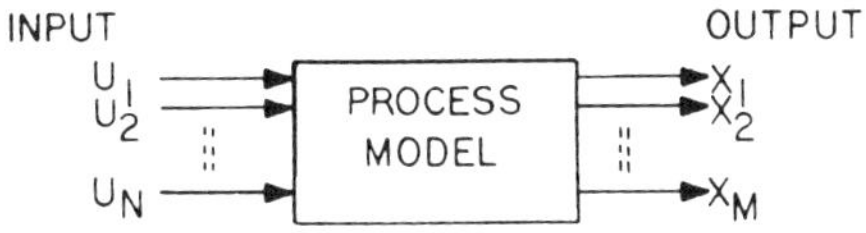

FIGURE 6.1 Functional representation of a process model showing input variables, u, (manipulatable variables), and output variables, x, (performance variables).

to the distillation process where x_{fi}, the feed composition, is a disturbance input. The overhead temperature T, column pressure P, distillate flow F_d, bottom flow F_b, distillate composition x_{di} (i = 1, 2, . . . n), and bottom composition x_{bi} (i = 1, 2, . . . n) can be considered as output. The block diagram representation of input/output for the distillation column can be simplified as shown in Fig. 6.3.

This input/output relationship for a given process generally can be considered as the cause-effect relationship of the process and is the most important relationship for understanding the actual manufacturing process. The responses deviated from the desired responses can be corrected if the input variables are properly manipulated to cause a direct or indirect effect on the process response.

Because the cause and effect relationship is considered the most important factor in the actual manufacturing process, the mathematical models created to simulate the actual manufacturing process must reflect

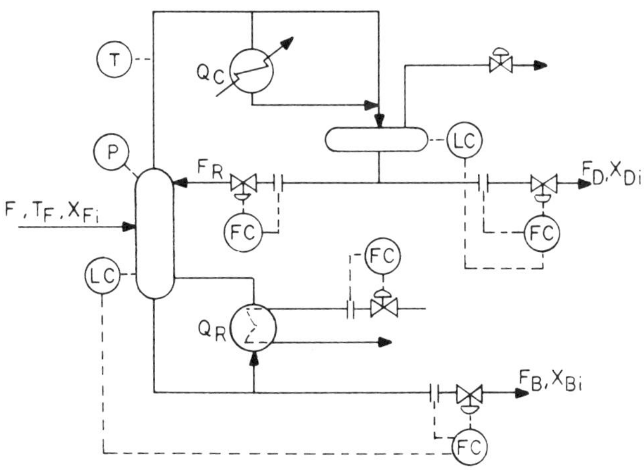

FIGURE 6.2 Simple process flow diagram of a distillation column showing control instrumentation.

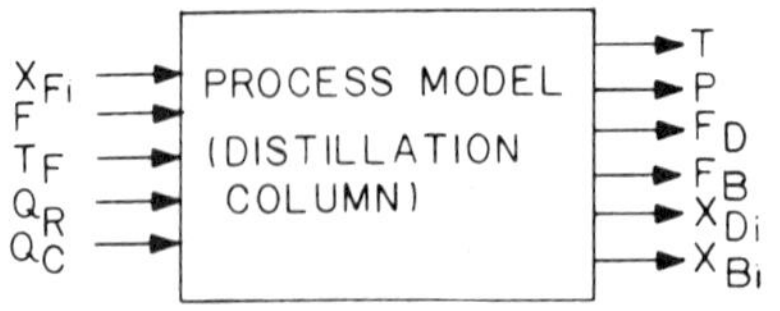

FIGURE 6.3 Process model representation of distillation column showing input (manipulatable) and output (performance) variables.

these input/output relationships. These relationships sometimes can be represented by the following mathematical expressions.

$$
\begin{aligned}
X_1 &= F_1(X_1, X_2, \ldots X_n, U_1, U_2, \ldots U_r) \\
X_2 &= F_1 (X_1, X_2, \ldots X_n, U_1, U_2, \ldots U_r) \\
&\vdots \\
X_n &= F_n(X_1, X_2, \ldots X_n, U_1, U_2, \ldots U_r)
\end{aligned}
\tag{6.1}
$$

Where X_i ($i = 1, 2, \ldots n$) represents the *ith* process output and U_j ($j = 1, 2, \ldots r$) represents the *jth* process input.

Both process input and process output may be represented by the material flow rate, energy flow rate, temperature, pressure, level, density or composition, etc., for typical industrial process. The functional relationship represented by $F_1, F_2, \ldots F_n$ are typically derived from the natural laws such as Equation of State, Conservation equations, and Transfer-rate equations.

6.2.2 Process Dynamics

The mathematical representation for input/output shown in Eq. (6.1) can mislead the actual response of the process in timing if one does not consider the actual dynamical behavior of the process. This can be illustrated from Eq. (6.1) by changing the input U_j from U_{j1} to U_{j2}. If the unique solution exists for Eq. (6.1), then the response of the process output X_i will be changed from X_{i1} to X_{i2}. Two major important questions are not answered.

1. How long it takes to reach X_{i2} from X_{i1} when U_j changes from U_{j1} to U_{j2} in the actual process.

2. What response trajectory the process output takes to reach X_{i2} from X_{i1}, when process input U_j changes from U_{j1} to U_{j2}. This is shown in Fig. 6.4.

Without this vital information contained in the model, the input/output relationship becomes unrealistic. Therefore, Eqs. (6.1) will be correct representations for instantaneous process where the output response to the input is instantaneous. If the Eq. (6.1) is to be used to represent a steady state, the person who creates the model must remember that he is utilizing an instantaneous mathematical representation for an extremely long dynamic process. In addition, the vital information mentioned in (1) and (2) is ignored.

Because of the difficulty in extracting the dynamic information from the steady state equation, the model created in a steady state manner is not so useful from the point of view of control system design. However, the steady state model is very useful for process design and steady state optimization. To approximate the actual operation more closely, a dynamic model is more realistic.

6.2.3 Identify the Manipulatable Inputs and Observable Output

As mentioned in the previous section, the relationship of the input/output in dynamic form is more realistic in the actual operation of a manufacturing process than in the steady state form. However, the inputs should be further classified to establish which inputs are actually manipulatable. Those inputs which are not manipulatable by control action generated by controllers or human operators should be considered as the disturbance inputs, as is shown in Fig. 6.5. The effect of these disturbance inputs can be very significant to counterbalance the manipulatable input.

For control system design, the disturbance input should be further classified as measurable disturbances or nonmeasurable disturbances.

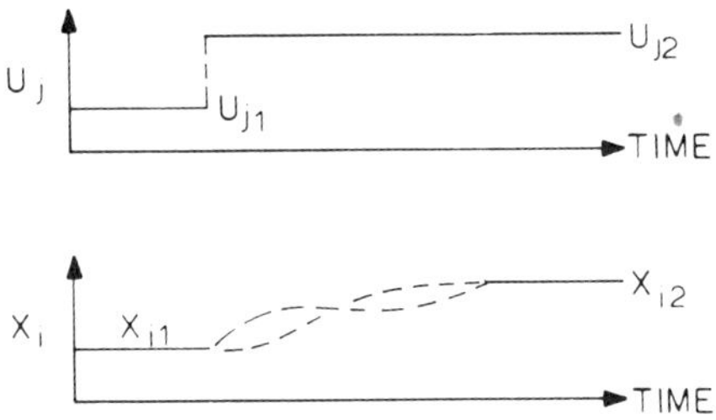

FIGURE 6.4 Timing diagram showing process output variable, x, response to step change in process input variable, u.

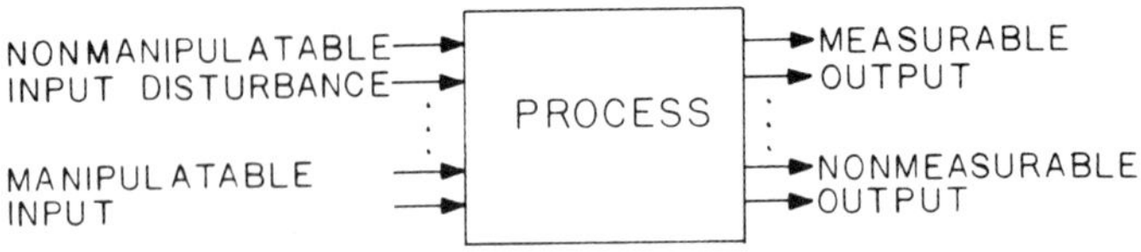

FIGURE 6.5 The process inputs are classified into manipulatable and nonmanipulatable input. The process outputs are classified into measurable and nonmeasurable output for process modeling.

Measurable disturbances can sometimes be compensated for by using either feedback or feedforward control actions. The output variable also should be further examined to determine if they are measurable. Some of the nonmeasurable output are observable through the use of a mathematical model of the known system input and output.

In most industrial applications, the direct measurement of the process variable has been utilized more in the actual control operation of the manufacturing process. The computed output from a process determined through the use of the mathematical models are not so popular in practical cases.

Because the mathematical model is not the actual process, the information generated from the model can be questionable. There are several reasons that can cause this mathematical model to behave unrealistically.

Examples are:

1. Assumptions required to create the equations for the models may be unreaslistic.
2. Model developer may be less experienced and, therefore, biased.
3. Sometimes too many unknowns exist due to lack of knowledge, and the model created is, therefore, incorrect.

Many industrial products have specified qualities which must be inspected before the product enters the market. The actual inspection test is usually a direct measurement. Therefore, an inferential type of computation through the model can create a credibility problem in the actual manufacturing environment; with the same reasoning, the process output variables such as temperature, pressure, flow rate, level, and composition, etc., are measured directly instead of computed through a mathematical model of the actual process.

6.3 TRANSFER FUNCTION

In many chemical plants, the dynamic behavior of the process can be expressed by a differential equation as shown in Eq. (6.2)

$$a_n \frac{d^n x}{dt^n} + a_{n-1} \frac{d^{n-1} x}{dt^{n-1}} + \ldots a_o x(t) = b \times u(t) \tag{6.2}$$

where x(t) is the output of the process or the state variable of the process. The process input variable, u(t), is a manipulatable input or a disturbance input. This differential equation can be transformed into an algebraic form called a transfer function. The transfer function usually is defined as the ratio of the Laplace transform of the output to input with all the initial conditions set to zeroes. If we take the Laplace transform of Eq. (6.2) and rearrange it, we have

$$\frac{X(s)}{U(s)} = \frac{b}{(a_n s^n + a_{n-1} s^{n-1} + \ldots + a_o)} \tag{6.3}$$

where

$$X(s) = \int_o^\infty x(t)e^{-st}\, dt \triangleq L[x(t)] \tag{6.4}$$

$$U(s) = \int_o^\infty u(t)e^{-st}\, dt \triangleq L[u(t)] \tag{6.5}$$

$$\frac{X(s)}{U(s)} = G(s) = \text{Transfer function} \tag{6.6}$$

The important aspects of this transfer function, G(s), is that it represents the dynamical behavior of the process in the complex domain, "s", instead of time domain, "t", and reduces the ordinary differential equation to an algebraic equation. This transformation in many situations can simplify the analysis and design for the linear control system.

6.4 MODEL TYPES

The mathematical model created from the actual process is usually derived from the combination of material, energy, momentum, or other equations derived from physical laws.

Depending on the model designer's preference, these models can sometimes be represented by the partial differential equation. This type of model is called a distributed parameter system. The dependent variables, which are process variables in most cases, are a function of the position and time in a given coordinate system. This means that the process variables are location and time dependent. Therefore, the sensor location for process variable measurement is an important consideration for the distributed parameter system.

Often the process model can be represented by the ordinary differential equation and this type of model is called a lumped parameter system. In this case, the process variables are not dependent on the position of the coordinate system. They are a function of time only. In many situations, the process model may be represented by nonlinear instead of linear differential equations or linear partial differential equations. The model designer should not impose too many assumptions on the physical process so as to modify the model in such a way that an analytical solution is possible. If the model is created in this manner, the output response to the input tends to be biased. The created model will not be useful in the actual process control application.

Most processes are subjected to the external or internal disturbances, which are random in nature. The characteristics of the disturbance may be modeled through the use of statistical methods. Process modeling and control design are segregated with a dotted partition line as shown in Fig. 6.6.

Functionally, the measured variable, z, is the function of the output, x, and noise, n. That is

$$z = g(x, n) \tag{6.7}$$

The output, x, is a function of disturbance, d, and input, u, which may be a manipulatable input. That is

$$x = h(u, d) \tag{6.8}$$

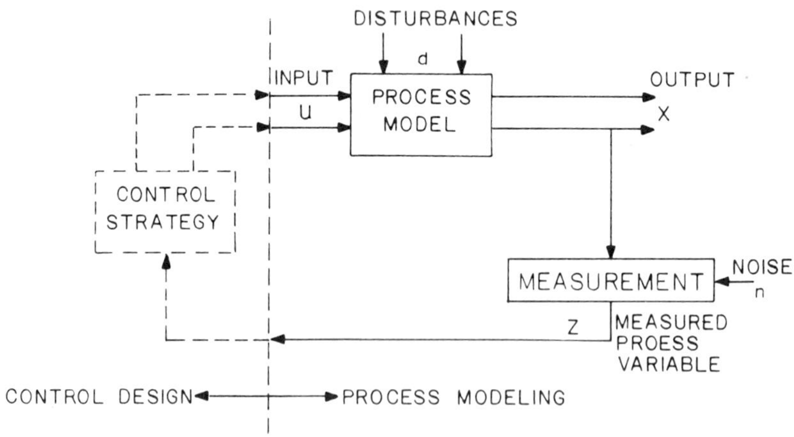

FIGURE 6.6 Simple feedback control is used to illustrate the relationship of process modeling and control strategy design.

where h represents the input/output relationship of the process model, and g represents the input/output relationship of the measurement device.

6.5 INCREMENTAL MODEL

The incremental model defined here is created from a procedure which imposes a small perturbation of the input variables for a process (7). The incremental output response due to incremental change of input can be estimated through the use of experimental methods or theoretical methods (12,13).

Experimentally, this model can be created from the actual plant in operation by using the pulse test method. By applying a small perturbation to the manipulated input and recording the output response in a synchronized manner, a complete set of input/output data can be collected. This is shown schematically in Fig. 6.7.

A step input is applied to the process and the output data are recorded. If we plot the input and output data, the input/output wave form can be observed. This is shown in Fig. 6.8. The output data collected for 0_1, 0_2, and I_1 as increment inputs, can be utilized to construct a meaningful model based on the physical or chemical phenomenon.

The output 0_1, or 0_2 is measured in an incremental manner from the steady state. If the model structure is defined, the problem is to find a parameter set which can be used in the model to produce the output response due to a given input change. This is called the parameter identification problem and will be discussed in detail later (14,15).

However, if the model structure is not defined, it is desirable to construct a model structure based on physical or chemical properties or laws. Without the physical or chemical concepts needed for creation of the model structure, any model created from pure input/output data obtained through a statistical method may generate erroneous results. The failure in the created model, such as from a typical regression model, is not because of the statistical method itself. This is because

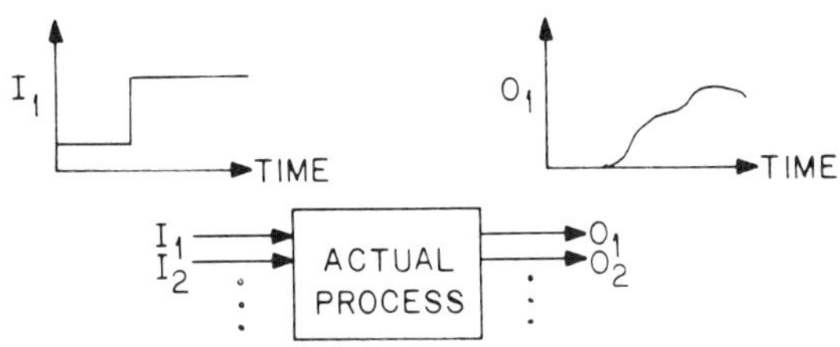

FIGURE 6.7 Open loop pulse test of the actual process using step type input.

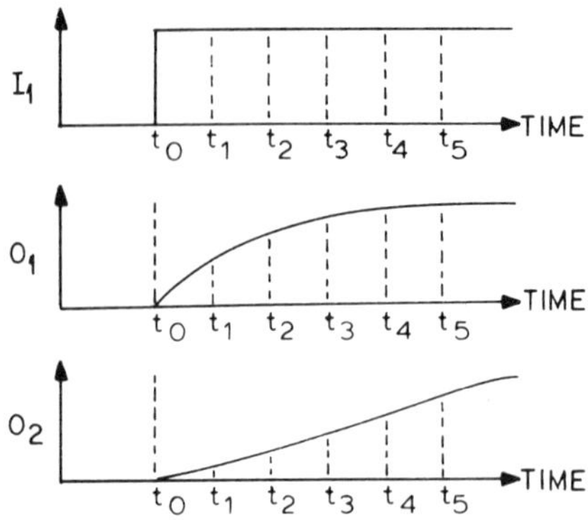

FIGURE 6.8 Input/output data are plotted with the recorded data synchronized. The input is a step type wave form. The measured output can be multiple output with different response characteristics.

the model designer did not utilize common sense in physics or chemistry to consider the model structure design. It is the responsibility of the model designer to ensure that the created model is applicable in practice.

It will be shown in the next section that an incremental model can be obtained from the linearized model for the process which can be analytically expressed in terms of linear or nonlinear differential equations. In this case, the structure of the model is defined. The parameters of the process can be obtained analytically by evaluating the function of the coefficient in the linearized equation without resorting to the experimental pulse test. However, for actual application of the model, it is recommended that the experimental test of the actual operation should be conducted to evaluate the validity of the model.

6.6 LINEARIZED MODEL

In most industrial processes, linear control behavior exists within a limited range of the operation. Generally speaking, the behavior of the nonlinear system can be approximated through the use of the piecewise linear approximation method. This concept is very important for control algorithm design, especially when applied to process regulation.

Assume that a typical dynamical process which is represented by a nonlinear differential equation of the following type is considered

$$\frac{dx}{dt} = f(x, u) \tag{6.9}$$

The variables x and u are expressed in an absolute position value with zero being the absolute reference position. Physically, this variable can represent a pressure, temperature, flow rate, or concentration, etc. This type of model will be defined here as an absolute model or position model. The feedback control designed for this system can be represented as shown in Fig. 6.9.

Let x_s be the solution trajectory for a given control input, u_s, satisfying the following equation

$$\frac{dx_s}{dt} = f(x_s, u_s) \tag{6.10}$$

Suppose the input, u, is deviating slightly from u_s, then the output, x, may be deviated also slightly from x_s, then

$$x = x_s + \Delta x \tag{6.11}$$

$$u = u_s + \Delta u \tag{6.12}$$

where Δx and Δu are incremental values of deviation from x_s and u_s, respectively. By subtracting Eq. (6.10) from Eq. (6.9), we have

$$\frac{dx}{dt} - \frac{dx_s}{dt} = f(x, u) - f(x_s, u_s) \tag{6.13}$$

Applying the Taylor series approximation to f(x, u), Eq. (6.13) can be expressed as follows by neglecting the higher order terms.

$$\frac{d\Delta x}{dt} = \frac{\partial f(x_s, u_s)}{\partial x} \Delta x + \frac{\partial f(x_s, u_s)}{\partial u} \Delta u \tag{6.14}$$

This is a linearized differential equation which is called a linearized model or imcremental model. This model represents an approximated

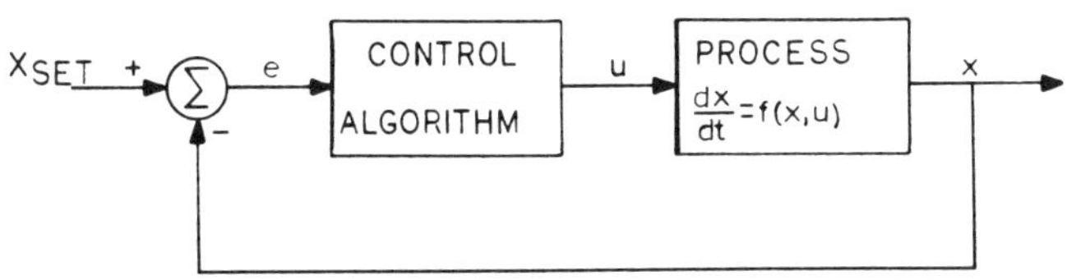

FIGURE 6.9 Feedback control of a nonlinear process is illustrated. This model will be used to illustrate the applicability of linearized models for regulation type control

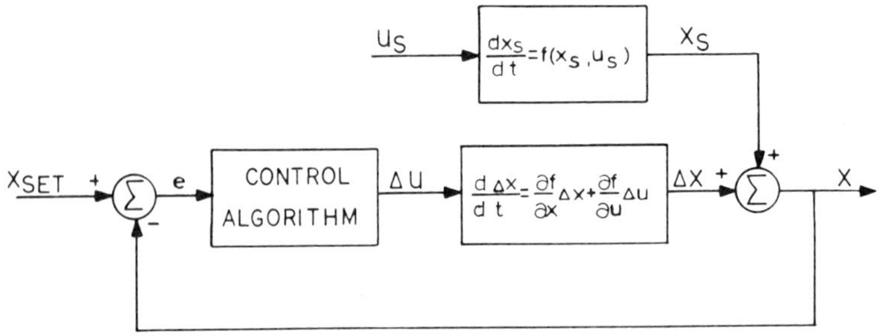

FIGURE 6.10 Typical linearized model for feedback control. This model is obtained from the original nonlinear process equation ($dx/dt = f(x, u)$).

behavior of the process in the neighborhood of the operating level at (x_s, u_s).

Let

$$A = \frac{\partial f\,(x_s, u_s)}{\partial x} \tag{6.15}$$

$$B = \frac{\partial f\,(x_s, u_s)}{\partial u} \tag{6.16}$$

The new coefficients A and B are functions of x_s and u_s. If the function $f\,(x_s, u_s)$ is linear in x_s and u_s, then the variables A and B are constants.

The model shown in Eq. (6.14) is extremely important, because most process controls in practice are based on this type of incremental model to design the feedback control algorithm. The linearized feedback control system is shown in Fig. 6.10.

6.7 DISCRETE STOCHASTIC MODEL

For computer control of the industrial process, the discrete representation of the process model in vector difference equations is convenient. Usually the system model is represented by the following vector equations.

The process model is

$$x(k+1) = \phi(k+1, k)\,x(k) + \Gamma(k+1, k)\,w(k) + \Psi(k+1, k)\,u(k) \tag{6.17}$$

where

$$k = 0, 1, 2, \ldots \text{ (time sequence)}$$

$$x = n - \text{dimensional vector for state variables}$$

$$\phi(k + 1, k) = n \times n \text{ state transition matrix}$$

$$w = p - \text{dimensional vector for disturbance}$$

$$\Gamma(k + 1, k) = n \times p \text{ disturbance transition matrix}$$

$$u = r\text{-dimensional vector for manipulatable input}$$

$$\Psi(k + 1, k) = n \times r \text{ control transition matrix}$$

The measurement model is

$$z(k + 1) = H(k + 1)\, x\,(k + 1) + v(k + 1) \tag{6.18}$$

where

$$z = m - \text{dimensional vector for measurement}$$

$$H = m \times n \text{ measurement matrix}$$

$$v = m - \text{dimensional vector for noise or measurement error}$$

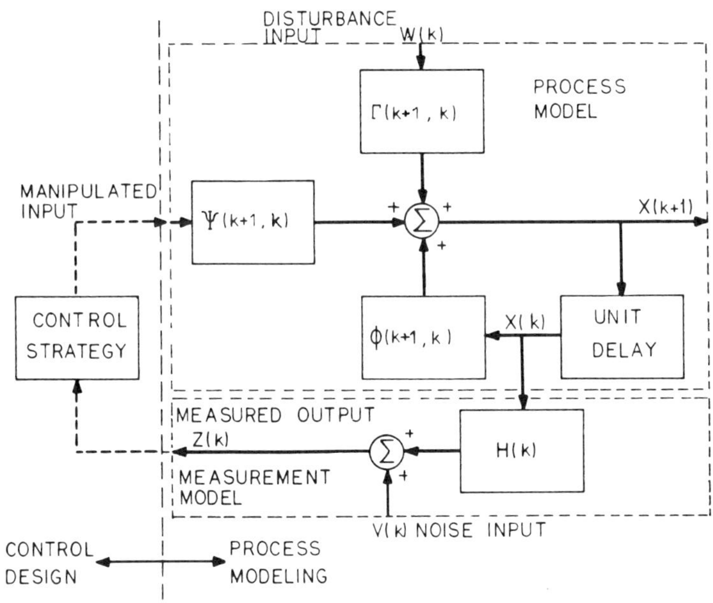

FIGURE 6.11 A discrete stochastic process model, represented with a matrix coefficient is illustrated. These matrix coefficients can be determined by using the parameter identification method.

The block diagram representation of the discrete stochastic model is shown in Fig. 6.11. The closed-loop system in Fig. 6.11 is subdivided into two major tasks. One is the process modeling task, and the other is the control design task.

6.8 PULSE TESTING TECHNIQUE

The incremental dynamic model is extremely important for computer control applications, especially for existing plants. Theoretically, the dynamic model can be created for the plant by subdividing the entire plant into smaller units. Each unit or region may be further analyzed to obtain a set of equations governing the operating behavior of the process. The complexity of the resulting equations can sometimes be so enormous that even modern computers may not be able to solve the problem in realtime.

Therefore, a simplified model, which is accurate enough for real-time computer control application, is extremely important. In many situations, this simplified model can be obtained experimentally by applying the pulse testing technique, which usually lasts for several minutes or hours, depending on the process dynamic characteristics. The input wave forms are usually step type, pulse type, or sinusoidal type. Each type of input wave form has its advantages and disadvantages in practical application.

6.8.1 Step Type Input

The step type input is shown in Fig. 6.12. This type of input is simple to apply in practice. The input variable is not necessarily the valve position. If the input variable is controlled by a closed-loop controller, the setpoint can be used as a manipulatable input which is popular in supervisory control. In this case, the setpoint change can be considered as a step type input.

The resulting transfer function, $G_r(s)$, is the combination of the closed-loop transfer function, $G_c(s)$, and the process transfer function, $G_p(s)$. This is shown in Fig. 6.13. The closed-loop transfer function, $G_c(s)$ is the ratio of the controlled input variable, u, to the setpoint of the input variable, u_{set}.

The process transfer function, $G_p(s)$, is the ratio of the output, x, to the input, u. This transfer function, $G_p(s)$, can be obtained if $G_r(s)$ and $G_c(s)$ are identified.

$$G_p(s) = \frac{G_r(s)}{G_s(s)} \tag{6.19}$$

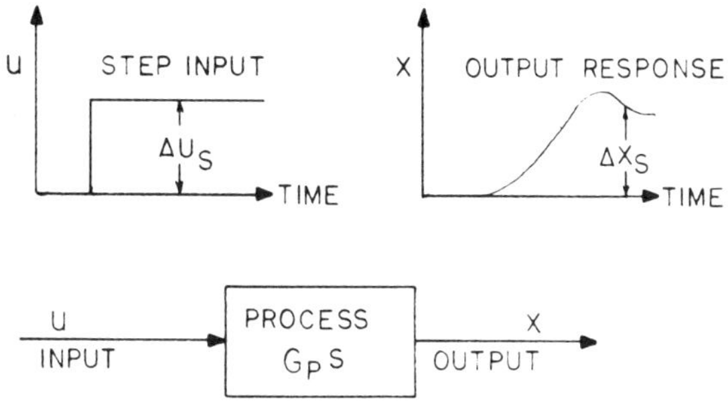

FIGURE 6.12 A step type input is used for a pulse test of the process to be modeled. This approach is relatively simple and practical to apply in practice.

Because of the nature of the input wave form, the response of the process, x, in low frequency range can be identified relatively easily. The steady state process gain parameters, k_p, at the given operating level can be obtained relatively accurately. The process gain parameter, k_p, is obtained by computing the following incremental input/output ratio:

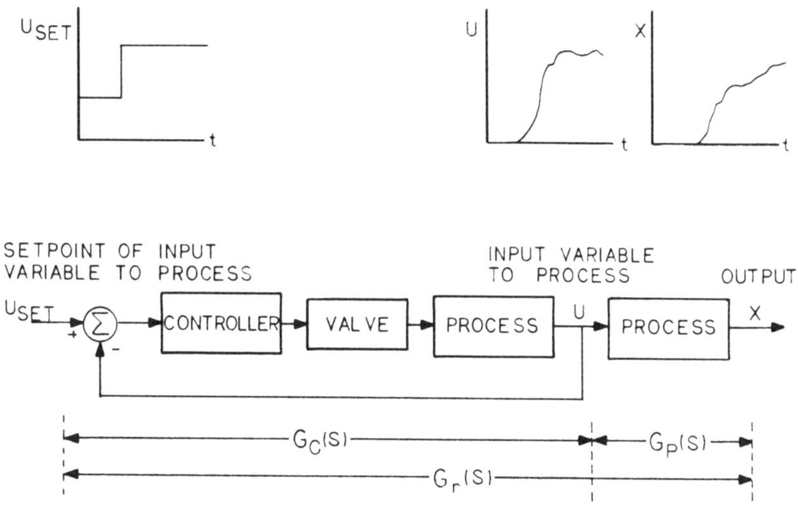

FIGURE 6.13 If the step input is applied at the setpoint of a controller, the output response is affected by the closed-loop behavior of the input process variable, u. For supervisory control, this behavior should be taken into account for process modeling.

$$k_p = \frac{\Delta x_{ss}}{\Delta u_{ss}} \tag{6.20}$$

where

Δx_{ss} = steady state incremental output

Δu_{ss} = steady state incremental input

If the process dynamics are first order, which is popular in chemical processes, the transfer function of the process, $G_p(s)$, can be expressed as

$$G(s)_p = \frac{k_p}{\tau s + 1} \tag{6.21}$$

where

τ = the time constant of process

k_p = the process gain parameter

The process dynamics characteristic parameters, such as τ and k_p for the first order process response, can be determined graphically. This is shown in Fig. 6.14. With modern distributed digital control systems, and with a central operator console and computer, any real time data accessible by the distributed control system can be displayed in a graphic manner on the console as shown in Fig. 6.8. The process

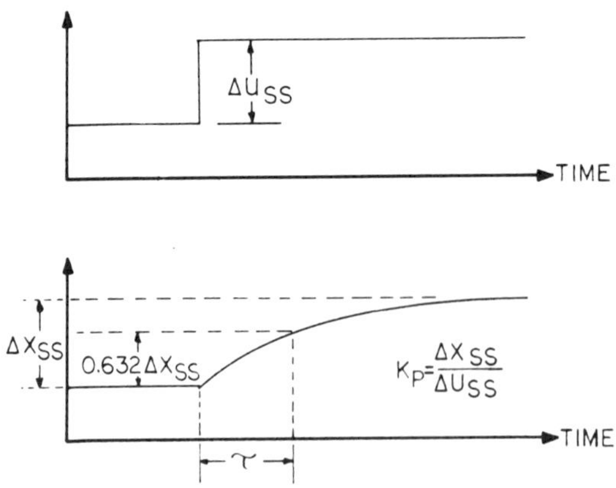

FIGURE 6.14 A simple first order process response to a step input is illustrated to determine the process gain k_p and time constant, τ.

characteristic parameters can be determined by visual inspection. If the parameters are required to be determined accurately, the parameter identification program can be used. This identification algorithm will be discussed later.

6.8.2 Pulse Type Input

The pulse type input is any input which has a closed wave form with respect to the steady state reference level as shown in Fig. 6.15. This type of wave form is a popular practical method to obtain dynamic information of the process. The frequency of the input wave form can cover a relatively wide range. Therefore, the dynamic parameters of the process can be obtained by applying the Fourier Transform method (12).

6.8.3 Sinusoidal Type Input

Sinusoidal type input is shown in Fig. 6.16. Theoretically, this is a simple method by which to obtain a wide range of the frequency responses by applying a wide range of input frequencies. However, in practice this can be impractical and very costly for use in the average process application. The reason is that, in order to obtain reasonable dynamic information on the model parameter for the frequency range of interest, the test period may be prohibitively long during normal plant operation.

The number of cycles of the sinusoidal input required to be applied to the process should be a sufficient number for sinusoidal output to reach the steady state. The testing period can be extremely long for a given frequency. The same type of testing should be conducted

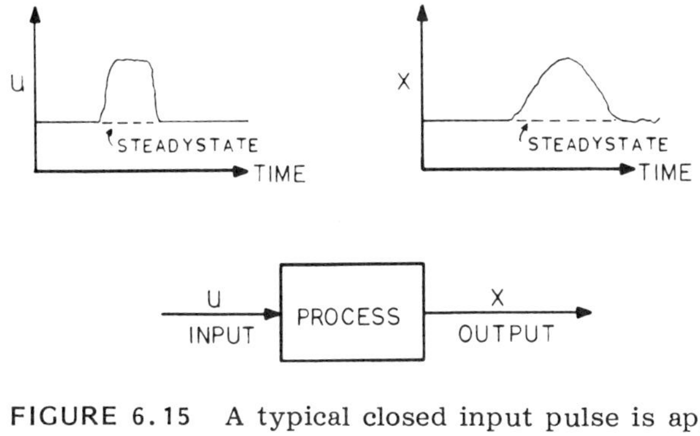

FIGURE 6.15 A typical closed input pulse is applied to the process. The output is recorded for model development.

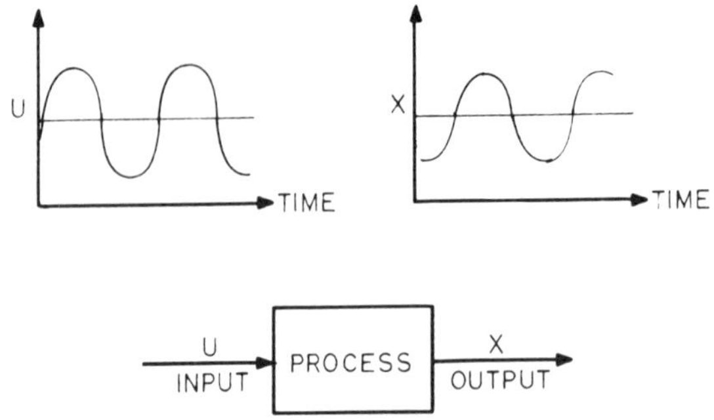

FIGURE 6.16 A typical sinusoidal wave form is applied to the input of the process. The output is recorded in a synchronized manner to determine the amplitude ratio and phase shift.

repeatedly for different frequencies in the range of interest. Therefore, the total testing period is too long to be practical.

6.8.4 Combination of Step Inputs

In actual application of the pulse testing procedure, a series of step inputs with different amplitudes, durations, or directions in any combination can be applied. This is shown in Fig. 6.17.

The input amplitude should be restricted to the range where an off-specification product should not be created. The duration of each step input should be long enough to extract major process parameters such as transportation delay, time constant, or process gain parameters. The direction of the step inputs should be changed to evaluate any nonlinear effect of the process. It will be shown in the control algorithm design that the parameter obtained from the actual test can be used as the actual tuning parameters for closed-loop

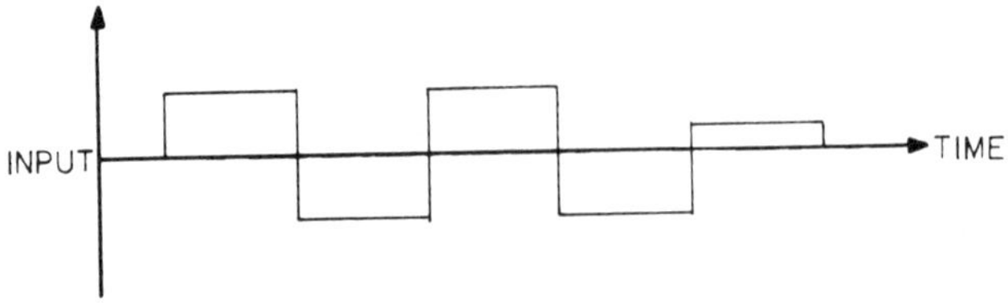

FIGURE 6.17 A combined step input in series is a practical and effective method to conduct the pulse test. This method can detect a nonlinear effect.

control implementation. This approach is practical for the actual implementation of the modern distributed digital system. Especially, as a computer is used for closed-loop control.

The time period required for an engineer to perform reasonably good controller tuning can be substantially reduced, because the tuning is accomplished in a more scientific manner than the seat-of-the-pants methodology which is typically practiced.

6.9 PROCESS PARAMETER IDENTIFICATION

The process dynamic model which can be described by a linear discrete system is considered. This type of mathematical representation of the process is often encountered in the process plant where a small perturbation method is applied. Especially, the sampled data control in supervisory control mode or direct digital control mode is applied to the process, the discrete dynamic model becomes very important for control strategy and algorithm design.

In general, a discrete dynamic equation can be written as

$$z(t) + a_1 z(t-1) + \ldots + a_n z(t-n) = b_1 u(t-1) + b_2 u(t-2) + \ldots + b_n u(t-n) \tag{6.22}$$

where $z(t-i)$ and $u(t-i)$ are the measured output and input values at sampled time $t - i$ ($i = 1, 2, \ldots n$). The process model parameters a_i and b_i are assumed constants which are to be determined from the set of observed input/output data. In this discussion only a least-squares fit method using sequential algorithm for the parameter estimate is presented.

Equation 6.22 can be written as

$$z(t) = h(t-1)\Phi + v(t) \tag{6.23}$$

where

$$\Phi = [-a_1, -a_2, \ldots, -a_n, b_1, \ldots b_n]^T$$

$$= 2n \times 1 \text{ vector}$$

$$h(t-1) = [z(t-1), z(t-2), \ldots, z(t-n), u(t-1), u(t-2), \ldots, u(t-n)]$$

$$= 1 \times 2n \text{ vector}$$

$$v(t) = \text{random noise}$$

Let's assume more sampled sets on input/output are observed, then

$$\begin{vmatrix} z(t+1) \\ z(t+2) \\ \vdots \\ z(t+n) \\ \vdots \\ z(t+m) \end{vmatrix} = \begin{vmatrix} h(t) \\ h(t+1) \\ \vdots \\ h(t+n-1) \\ \vdots \\ h(t+m-1) \end{vmatrix} \Phi + \begin{vmatrix} v(t+1) \\ v(t+2) \\ \vdots \\ v(t+n) \\ \vdots \\ v(t+m) \end{vmatrix} \tag{6.24}$$

This can be written as the following vector equation

$$Z = H\Phi + V \tag{6.25}$$

where

$$Z = \begin{vmatrix} z(t+1) \\ z(t+2) \\ \vdots \\ z(t+m) \end{vmatrix} = m \times 1 \text{ vector}$$

$$H = \begin{vmatrix} h(t) \\ h(t+1) \\ \vdots \\ h(t+m-1) \end{vmatrix} = m \times 2\,n \text{ matrix}$$

$$V = \begin{vmatrix} v(t+1) \\ \vdots \\ v(t+m) \end{vmatrix} = m \times 1 \text{ vector}$$

It is desired to obtain the best estimate of Φ denoted by $\hat{\Phi}$, such that the cost function, J, in quadratic form

$$J = \frac{1}{2}[Z - H\Phi]^T R^{-1} [Z - H\Phi] \tag{6.26}$$

is minimized, where R is weighting matrix. If R is selected to be a unit matrix, the result of the estimation is in a minimum error square sense.

The necessary conditions for a minimum are

$$\frac{\partial J}{\partial \Phi} = H^T R^{-1}(Z - H\Phi) = 0 \tag{6.27}$$

by which we mean that

$$\frac{\partial J}{\partial \Phi_i} = 0 \text{ for } i = 1, 2, \ldots 2n$$

where

$$\Phi = [\Phi_1, \Phi_2, \ldots \Phi_{2n}]^T = [-a_1, -a_2 \ldots -a_n, b_1 \ldots b_n]^T$$

The best estimate of Φ, $\hat{\Phi}$ becomes

$$\hat{\Phi} = (H^T R^{-1} H)^{-1} H^T R^{-1} Z \tag{6.28}$$

For least error square estimate,

$$R = \text{unit matrix}$$

Equation (6.28) reduces to the following

$$\hat{\Phi} = (H^T H)^{-1} H^T Z = P_m H^T Z \tag{6.29}$$

where

$$P_m^{-1} = H^T H$$

when a new sampled measurement at t + m + 1 is obtained. The best estimate of Φ for least error square is

$$\Phi_{m+1} = \left[\left(\frac{H}{h_m}\right)^T \left(\frac{H}{h_m}\right)\right]^{-1} \left[\frac{H}{h_m}\right]^T \left[\frac{Z}{z(t+m+1)}\right]$$

$$= P_{m+1} \left[\frac{H}{h_m}\right]^T \left[\frac{Z}{z(t+m+1)}\right]$$

$$= P_{m+1} \, [H^T Z + h_m^T z(t+m+1)] \tag{6.30}$$

where

$$P_{m+1}^{-1} = \left[\frac{H}{h_m}\right]^T \left[\frac{H}{h_m}\right]$$

$$= [\, H^T H + h_m^T \, h_m]$$

$$= [\, P_m^{-1} + h_m^T \, h_m] \qquad (6.31)$$

$$h_m = h(t + m) \qquad (6.32)$$

$$h_m^T = \text{transpose of } h_m$$

Applying a matrix inversion lemma to Eq. (6.31), we have

$$P_{m+1} = [P_m^{-1} + h_m^T \, h_m]^{-1}$$

$$= P_m - P_m h_m^T \, [\, h_m \, P_m \, h_m^T + 1]^{-} h_m P_m \qquad (6.33)$$

Substituting Eq. (6.33) into Eq. (6.30) and Φ_m for $\hat{\Phi}$, the following recursive algorithm is obtained.

$$\Phi_{m+1} = \Phi_m + P_m h_m^T \, [\, h_m P_m h_m^T + 1]^{-1}$$

$$[\, z(t + m + 1) - h_m \Phi_m] \qquad (6.34)$$

6.10 PROCEDURE FOR MODEL DEVELOPMENT

To conclude the process modeling, a simple flow diagram, as shown in Fig. 6.18, is utilized to illustrate the iterative procedure which has essentially no end if a satisfactory model cannot be obtained by the model developer. The effort, time, and cost can have no limit. Therefore, it is important that a reasonable limit should be imposed whether this limit is budget, time, or effort.

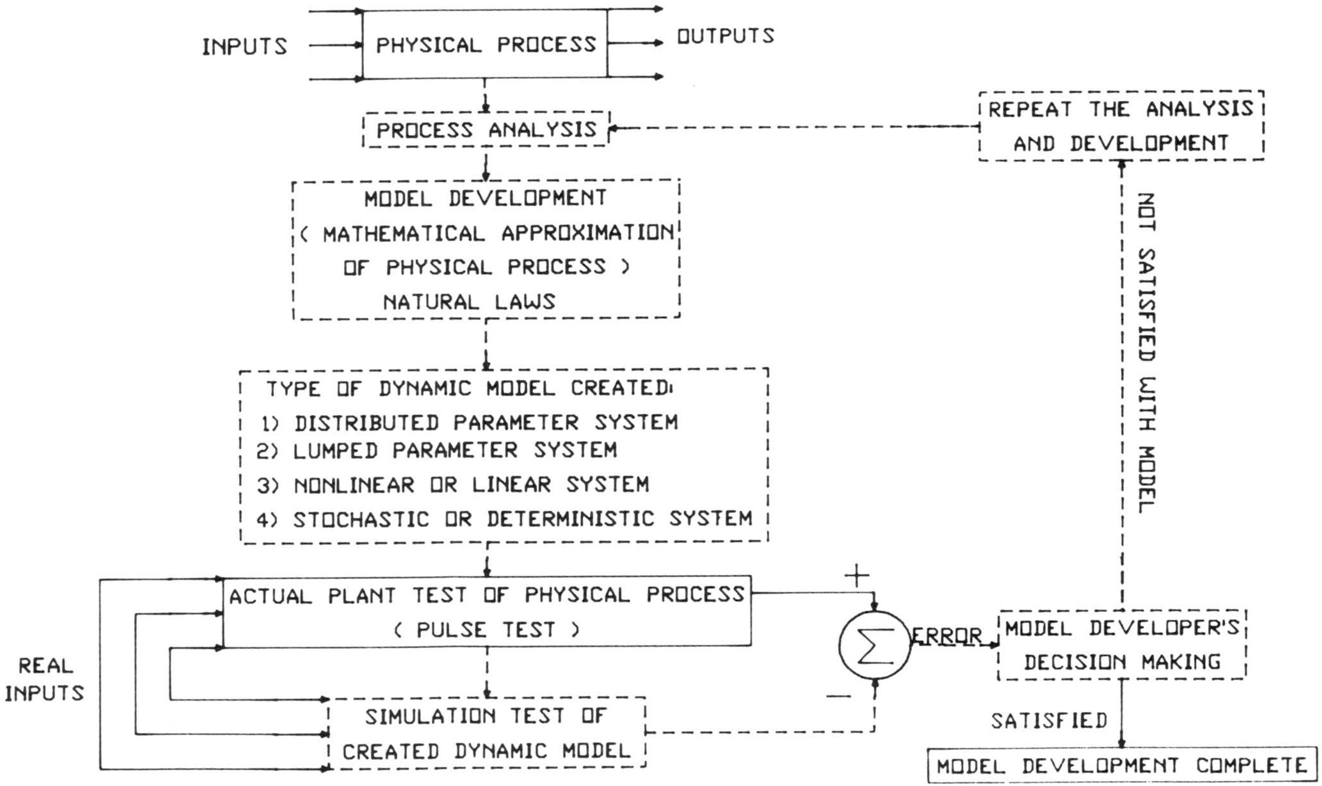

FIGURE 6.18 A schematic diagram for model development procedure is expressed in an iterative manner with an actual physical test involved.

REFERENCES

1. Yang, Wen-Jei and Masubuchi, Masami, *Dynamics for Process and System Control*, Gordon and Breach, New York, 1970.
2. Douglas, J. M., *Process Dynamics and Control Volume 1, Analysis of Dynamic System*, Prentice-Hall, Inc., Englewood Cliffs, New Jersey, 1972.
3. Smith, Cecil L., Pike, Ralph W., and Murrill, Paul W., *Formulation and Optimization of Mathematical Models*, International Textbook Company, Scranton, Pennsylvania, 1970.
4. Douglas, J. M., *Process Dynamics and Control Volume 2 Control System Synthesis*, Prentice-Hall, Inc., Englewood Cliffs, New Jersey, 1972.
5. Smith, Cecil L., *Digital Computer Process Control*, Intext Educational Publishers, Scranton, Pennsylvania, 1972.
6. Jamshidi, M., *Large-Scale Systems Modeling and Control*, North-Holland, New York, 1983.
7. Weber, Thomas W., *An Introduction to Process Dynamics and Control*, John Wiley & Sons, New York, 1973.
8. Mohillal, Rezso and Ferencz, Bela, *Chemical Process Dynamics*, Elsevier, New York, 1982.
9. Friedly, John C., *Dynamic Behavior of Processes*, Prentice-Hall, Inc., Englewood Cliffs, New Jersey, 1972.
10. Zoss, Leslie M., *Applied Instrumentation in the Process Industries Volume IV*, Gulf Publishing Company, Houston, Texas, 1979.
11. Robinson, E. R., *Time Dependent Chemical Processes*, John Wiley & Sons, New York, 1975.
12. Hougen, Joel O., *Measurements and Control Applications*, Instrument Society of America, Pittsburgh, Pennsylvania, 1979.
13. Gustavsson, Ivar, Survey of applications of identification in chemical and physical processes, Identification and System Parameter Estimation Part 1. *Proceedings of the 3rd IFAC Symposium, The Hague/Delft*, The Netherlands, June 12–15, 1973, pp. 67–85.
14. Mendel, Jerry J., *Discrete Technique of Parameter Estimation*, Marcel Dekker, Inc., New York, 1973.
15. Lin, C. S. and Tsai, T. H., Application of on-line process identification for adaptive control, Proceedings of the Eight Annual Conference on the Use of Digital Computers in Process Control, University of Tennessee, Knoxville, Tennessee, April, 1973, pp. 22–27.

7

Control Concept and Algorithm Design

7.1 INTRODUCTION

The derivation and application of control algorithms are discussed in this chapter. This is a favorite subject for many control experts, and has been the subject of many publications (1–3). The purpose of this chapter is not to echo the contents of these many publications, but to give engineers an explanation of a few frequently used control algorithms (4–6), and to discuss the concepts of their applications. In addition, a few not so frequently used multivariable control algorithms are introduced as examples for future development. These multivariable algorithms are derived from actual needs and installations, and they are much simpler than some of the algorithms developed from modern control theories.

Basically, there are two types of control levels in the calculations. One is the lower level regulatory control which defines the control action required to maintain the process variable to be as close as possible to the desired value or setpoint in the presence of the disturbances. The other types of calculations are the higher level control commands which are used to determine the "best" value of setpoints for several variables based on a predefined performance criteria. The term control algorithm generally refers to the first type of control calculation, and will be discussed in this chapter.

7.2 DISCUSSIONS OF CONTROL ALGORITHM DESIGN CONCEPTS

7.2.1 Feedback Control

Most existing process plants are operated under closed loop control utilizing a large number of control loops. The majority of these loops are the feedback control type. Feedback control can be considered as the basic type of control in the process industry because of its proven application and reliability.

Feedback control concepts can be explained according to the functional block diagram shown in Fig. 7.1. The purpose of the feedback controller in this diagram is to generate a controller output command signal, u, such that the error signal, e, approaches zero. When it does, the measured process variable should be approaching setpoint, y_{SET}.

Note that the true process output variable, x, is not usually equal to the measured variable, y, because of errors in the sensor and transmitter. The measured process variable, y, is the variable used for control.

Each basic functional block indicated in Fig. 7.1 can be considered as an information device or process. In addition, the functional block may need, in reality, to convert the input signal to the device or output signal from the device according to the physical property required for smooth interface between the functional blocks.

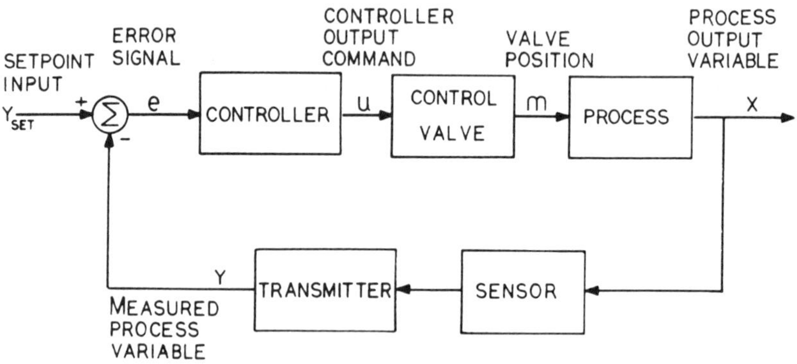

FIGURE 7.1 A functional block diagram for a typical feedback control loop is illustrated. The output variable, X, of the process is measured through a sensor and transmitted to the controller. The error signal, e, is generated in the controller with the setpoint value minus the measured process variable. The output of the controller, u, commands the control valve, which will manipulate the input to the process.

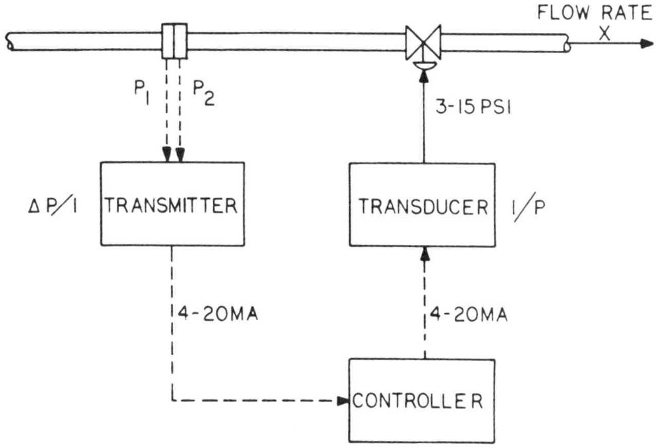

FIGURE 7.2 This flow control loop illustrates five major component blocks. Each block has a different physical property for input and output. The flow measurement is converted to the pressure differential, ΔP, which is transmitted to the typical controller with 4 to 20 ma signal. The output of the controller is generating a 4 to 20 ma signal for the transducer. The output of the transducer is a pressure signal (3 to 15 psi) for operating the control valve.

An example for a simple feedback control loop is illustrated in Figure 7.2, using an electronic device. The flow rate, x, shown in Fig. 7.2 is the process variable to be controlled. It is measured by the orifice plate type sensor to produce a differential pressure signal, Δp. The transmitter converts the differential pressure signal to a milliampere signal and sends it to the controller. The controller computes an output signal according to the selected control algorithm and sends an electronic output signal to the current-to-pressure transducer. The control valve receives the pressure signal and adjusts the valve position which affects the flow rate.

Mathematically, the control algorithm can be expressed as:

$$u = f(e) \tag{7.1}$$

where u is the output of the controller. The error signal, e, is usually defined as the difference between setpoint, x_{SET}, and measured process variable, x, (i.e., $e = x_{SET} - x$). The functional relationship, f, can be a linear or a nonlinear function or a logical function, as long as the control objectives are satisfied. The important aspect of the objectives is to manipulate the command signal, u, according to the desired control low or control algorithm so that the error signal, e, approaches zero. This type of controller is defined as a feedback controller because it only corrects the errors produced by a feedback signal as compared to a feedforward controller which is designed to compensate for disturbances before they affect the controlled variable. Feedback controllers are widely used because they are easily understood and can be used for a wide range of applications.

7.2.2 Incremental Algorithm and Position Algorithms

The incremental algorithm is also called a velocity algorithm (6,7). This algorithm is extremely important for the high level control command, such as an operator adjustment of the setpoint of an analog or digital controller, or computer supervisory control for setpoint commands. In actual operations, most of the human commands to adjust the manipulatable inputs can be interpreted in terms of the concept of the incremental algorithm.

An obvious example of incremental control can be experienced when driving a car by moving the steering wheel. The driver's eyes are the sensors for the car's position relative to the road. If the position of the car deviates from the guide line on the road, the driver usually corrects the deviation gradually by changing the steering wheels's position in an incremental manner. The correct steering position is accomplished by increasing or decreasing the angular position of the steering wheel from the past position regardless of the absolute position of the steering wheel. This is an extremely important concept for control. This means that to control the vehicle's position, x, the control command generated by the driver is not based on the absolute position value of the steering angle u(n), instead, the driver needs to know the approximate incremental adjustment, $\Delta u(n)$, required from the existing absolute position of the steering wheel whatever that position is.

This same concept of incremental control also can be applied to the process control environment. For digital control using sample-data control theory, the incremental control algorithm is usually expressed as

$$\Delta u_c(n) = f_v(e_n, e_{n-1}, \ldots) \tag{7.2}$$

where $e_n = x_{SET} - x(n)$, and $e_{n-1}, \ldots$, etc., are defined as the past errors. The value, n, represents the current sampled number which is related to the current sampling time, t_n, as shown in Fig. 7.3 and the following equation

$$t_n = nT \tag{7.3}$$

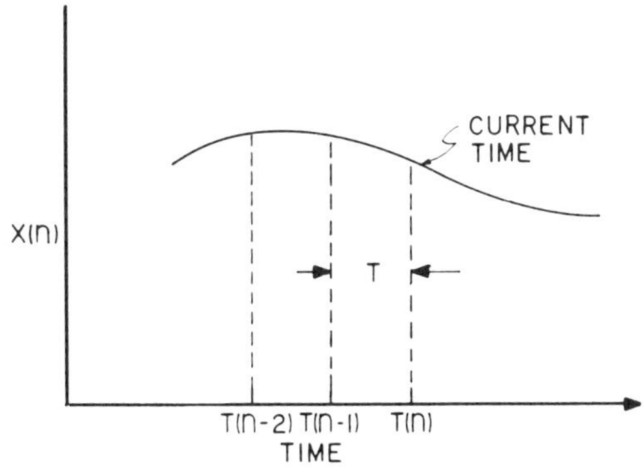

FIGURE 7.3 Typical sampled data values for continuous analog signal, x, are illustrated for sampling time T(n), T(n − 1), and T(n − 2) with sampling period T.

where T is the sampling period. For a position algorithm, the control command is usually expressed as

$$u(n) = f_p(e_n, e_{n-1}, \ldots) \tag{7.4}$$

The relation between position and incremental form is expressed by the following equation.

$$u(n) = u(n-1) + \Delta u_c(n) \tag{7.5}$$

Note that for a control loop whose output goes to a final control element, such as a valve, position output from a controller is usually used, because most valves can only be driven by position signals. (Some valves are driven by stepping motors which accepts only incremental signals.) This can be accomplished by using either Eq. (7.4) or Eq. (7.5). For most computer control loops, incremental algorithms are much more desirable than position algorithms because the output tracking and setpoint initialization functions using the incremental algorithm are much simpler. Figure 7.4 shows a cascade type control loop implemented by using a computer system. When the supervisory loop is turned on, it is sometimes necessary to have a "bumpless transfer" which will eliminate a disturbance to the existing flow controller setpoint.

The discrete form of PID algorithm in position form is as follows:

$$u(n) = k_p \left[e_n + \frac{T}{T_I}(\text{Sum }(n-1) + e_n) + \frac{T_D}{T}(e_n - e_{n-1})\right] \tag{7.6}$$

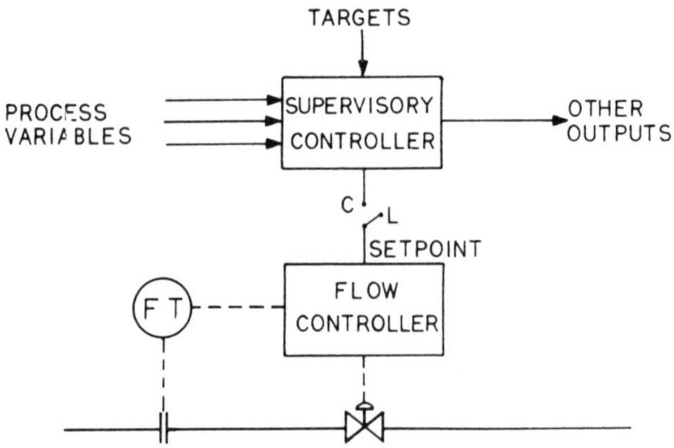

FIGURE 7.4 The cascade control is implemented with a supervisory computer system to illustrate the similarity of the cascade control with the analog control. The bumpless transfer of the setpoint from the supervisory computer is much more straightforward to implement than for the analog controller.

where

T = the sampling period

T_I = the integral time constant or reset time

T_D = the derivative time constant

k_p = the proportional gain constant

Sum (n − 1) = the result of the summation of the past errors.

This algorithm has been widely used for both simple control loops and cascade control loops. The initialization of this algorithm requires a special calculation because the Sum (n − 1) term will act as a bias term at the control initialization time.

To explain this initialization calculation, consider a cascade outer loop using a position PID algorithm shown in Eq. (7.6). When a position algorithm is used, automatic tracking and initialization functions must be implemented. For a single-variable controller, such as a cascade outer loop, tracking and initialization is not difficult to implement. However, for a supervisory controller using more than one process variable, it could be a complicated task to design tracking and initialization functions using the position algorithm. Tracking and initialization becomes much simpler when incremental algorithms are used. Basically, all incremental outputs can be set to zero to eliminate the initial "bump", and the target values (or set-point) may or may not be initialized depending on the control strategy. Further discussions on initialization will be made with control algorithm discussions.

7.3 FREQUENTLY USED CONTROL ALGORITHMS

7.3.1 Simple Feedback Control Algorithms (Position and Velocity)

There are several linear feedback control algorithms used in the process industries for computer control. Many control algorithms are direct translations from analog controllers. The most popular control algorithm combines the proportional, integral, and derivative (PID) functions. This is shown in Figure 7.5. The initialization calculations for this loop are

$$e_n = 0 \tag{7.7}$$

$$e_{n-1} = 0 \tag{7.8}$$

$$\text{Sum}\ (n-1) = \frac{T_I u\ (n-1)}{K_p T} \tag{7.9}$$

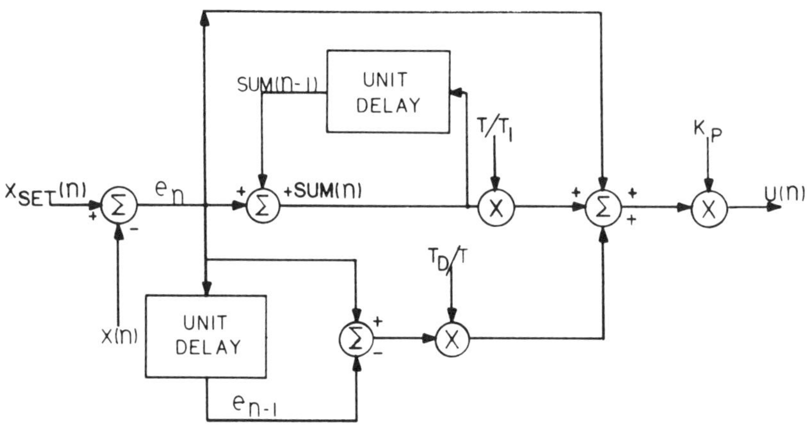

FIGURE 7.5 A discrete position form of a PID algorithm is illustrated. The error signal, e(n), is processed using proportional, integral, and derivative modes. The output, u(n), is the product of the proportional constant, K_p, and the combined result of the three modes.

$$x_{set}(n) = x(n) = \text{Process variable for outer loop} \tag{7.10}$$

$$u(n - 1) = \text{inner loop set point feedback (supervisory)} \tag{7.11}$$

The value, Sum (n − 1), is calculated from Eq. (7.9). This initialization algorithm for position form is an extra equation which needs to be implemented in conjunction with the PID algorithm in position form. The discrete form of the PID incremental (or velocity) algorithm is shown in Fig. 7.6 and can be expressed as follows:

$$\Delta u(n) = k_p \left[(e_n - e_{n-1}) + \frac{Te_n}{T_I} + \frac{T_d (e_n - 2e_{n-1} + e_{n-2})}{T} \right] \tag{7.12}$$

The advantage of this algorithm is that bumpless transfer can be accomplished without performing setpoint tracking when the lower loop control mode is changed from local automatic to cascade. The initialization calculation for the outer loop control algorithm are:

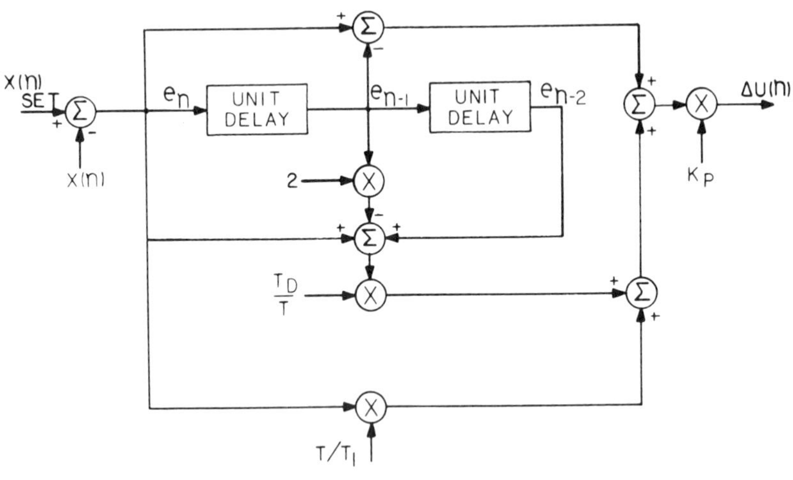

FIGURE 7.6 A discrete PID algorithm in velocity (incremental) form is illustrated. This algorithm is integrated to generate a position type output. The velocity output algorithm is very important for multi-input processes.

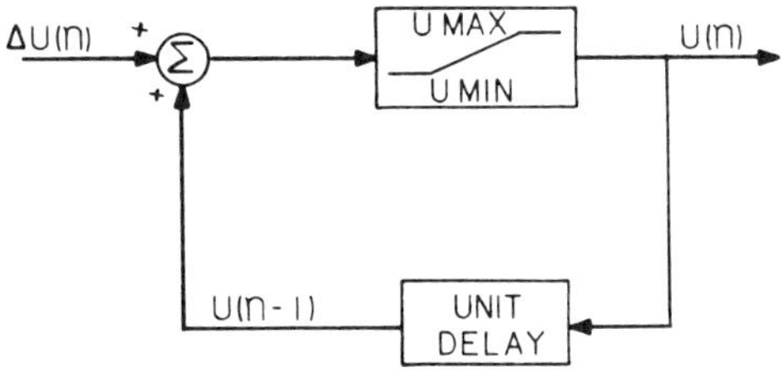

FIGURE 7.7 Integral control algorithm with saturation limits included to prevent the integral windup. This algorithm is used for positioning an output with incremental inputs. Usually, a position output of the PID algorithm is generated with the incremental output of a PID algorithm plus integral control with an antiwindup algorithm combined.

$$e_n = 0 \tag{7.13}$$

$$e_{n-1} = 0 \tag{7.14}$$

$$e_{n-2} = 0 \tag{7.15}$$

$$\Delta u(n) = 0 \tag{7.16}$$

$$x_{set}(n) = x(n) = \text{outer loop process variable} \tag{7.17}$$

When the inner loop's setpoint is in position form, it is necessary to integrate the output of the velocity algorithm of the outer loop. This is shown in Eq. (7.5), and in Fig. 7.7.

The term, U_{max}, is the maximum value of allowable innerloop setpoint. The term, U_{min}, is the minimum value of allowable inner loop setpoint. Both allowable values can be physical limits or process limits, depending on the applications. The nonlinear block introduced in Fig. 7.7 is to prevent the integral windup.

The integration function shown in Fig. 7.7 does not have to be part of the cascade outerloop. In fact, it is more desirable to implement that function as part of the inner loop. This allows the inner loop to accept incremental values from the outer loop and eliminates the need for transferring the inner loop's last setpoint value to the outer loop.

7.3.2 The Application of Cascade Control

Cascade controllers are widely used in process plants to control slow processes. This type of controller is used when intermediate measurements can be made between the controlled variable and the final control elements.

As an example, a furnace outlet temperature control function is selected for discussion. Two approaches are shown in Figs. 7.8 and 7.9. The difference between the two is that the first configuration is based on a simple feedback controller and the other is based upon a cascade controller. In these examples, six major disturbance sources are assumed as shown in Fig. 7.10.

Each of these disturbances can cause a change in the outlet temperature. When a simple feedback temperature controller is applied directly to the fuel valve position as shown in Fig. 7.8, the influence to the temperature due to all disturbances will be detected by the temperature sensor and corrected by manipulating the fuel flow valve position. When a flow control loop is used as an inner loop control as shown in Fig. 7.10, and the flow setpoint value is the absolute value of the output from the temperature controller, the influence to the temperature due to fuel pressure disturbances at locations (1) and (2) can be compensated for before they influence temperature measurements. For a process where the fuel header pressure varies significantly, this type of cascade controller can improve the temperature control performance substantially. However, disturbances (3), (4), (5), and (6) still exist. In this case, feedforward control technique can be employed in conjunction with the feedback to actuate the fuel flow setpoint.

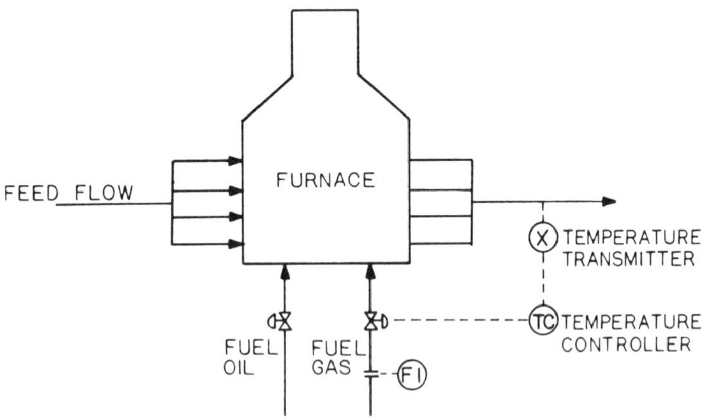

FIGURE 7.8 An example of a simple feedback control applied to the furnace outlet temperature control is illustrated. The output of the temperature controller is directly applied to the control valve of the fuel gas flow.

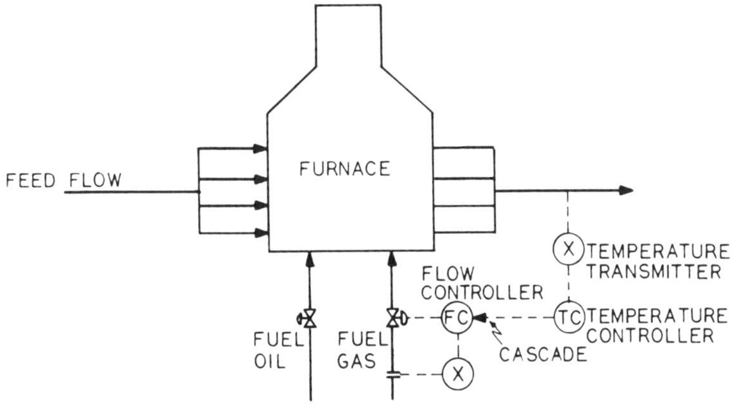

FIGURE 7.9 An example of cascade feedback control applied to the furnace outlet temperature control is illustrated. The inner loop is a fuel gas control loop. The setpoint of the fuel gas controller is generated from the output of the temperature controller.

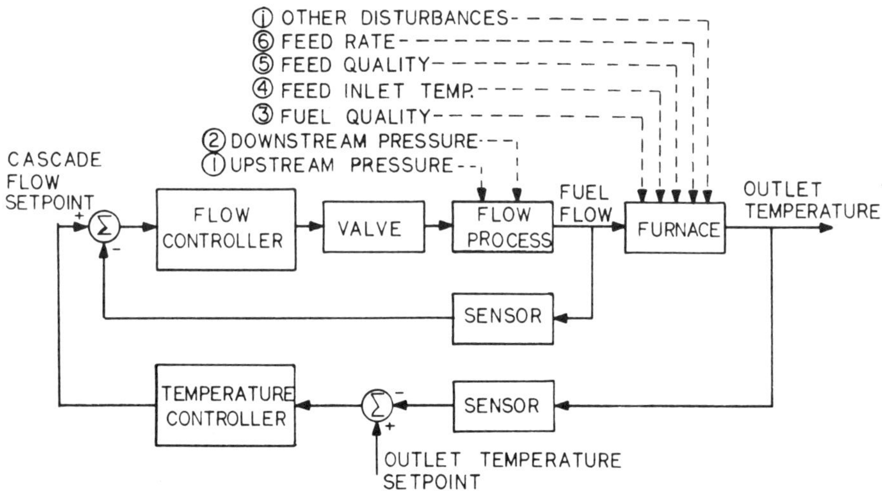

FIGURE 7.10 The block diagram of cascade control for a furnace outlet temperature with several disturbance sources is illustrated. The influence of these disturbances, relative to the inner and outer loops, is also classified to indicate the location of a disturbance input.

7.3.3 Feedforward Control

When the process output, x, shown in Fig. 7.11 is influenced by a measurable disturbance input, w, it is sometimes desirable to design a feedforward control function G_f, to actuate the manipulatable input in order to compensate for this undesirable influence from disturbances.

Assuming that the transfer function of the process is $G_p(s)$ and the transfer function of the process output to the disturbance input is $G_d(s)$, the simple feedforward control design is to determine the algorithm of transfer function G_f, such that

$$G_f G_p \times w + G_d \times w = 0 \tag{7.18}$$

provided that the disturbance, w, can be measured. The transfer function, G_f, is obtained by solving Eq. (7.18).

$$G_f = -\frac{G_d(s)}{G_p(s)} \tag{7.19}$$

This transfer function, G_f, may or may not be physically realizable. If the functions, $G_d(s)$ and $G_p(s)$, are first order processes which are represented by

$$G_d(s) = \frac{k_d}{T_d s + 1} \tag{7.20}$$

$$G_p(s) = \frac{k_p}{T_p s + 1} \tag{7.21}$$

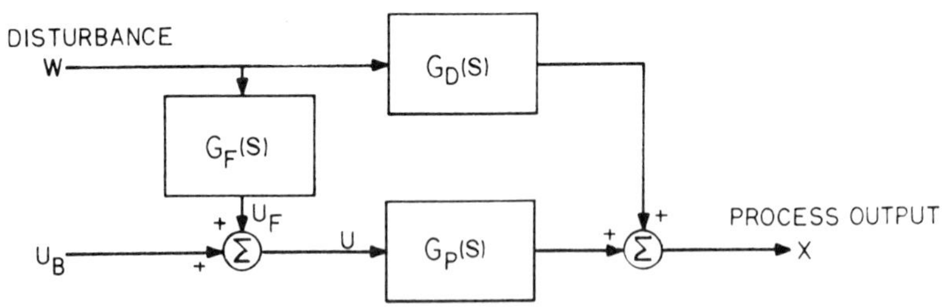

FIGURE 7.11 Block diagram representation of simple feedforward control is illustrated. The process transfer function is represented by $G_p(s)$. The disturbance transfer function is represented by $G_d(s)$. The design feedforward control transfer function is represented by $G_f(s)$.

The transfer function for feedforward control becomes

$$G_f(s) = -\frac{k_d}{k_p}\frac{T_p s + 1}{T_d s + 1} \tag{7.22}$$

This is a popular lead-lag transfer function implemented in both conventional analog and modern digital controllers. The discrete control algorithm for this lead-lag transfer function is

$$u'_f(n) = w(n) + \frac{T_p - T_d}{T_d}(w(n) - y(n)) \tag{7.23}$$

where

$$y(n + 1) = y(n) + \frac{T}{T_d}(w(n) - y(n)) \tag{7.24}$$

and

$$u_f(n) = -\frac{k_d}{k_p}u'_f(n) \tag{7.25}$$

When the output of the feedforward algorithm and the output of the feedback algorithm are combined to form a new output, u, the position form of the manipulatable input variable, u, becomes

$$u = u_{fb} + u_{ff} \tag{7.26}$$

This is shown in Fig. 7.12.

If the sensor of the feedforward loop has failed, the feedforward control signal, u_{ff}, will disappear. This will create a step change in the process input, u, as can be seen from Eq. (7.26). To prevent the unnecessary change in the manipulatable input, u, at this instant, the feedback controller output, u_{fb}, should be reinitialized to the past value of the manipulatable input, u(n − 1), to ensure that the bumpless transfer is realized in the presence of sensor failure.

When the feedforward loop in position form is to be connected with the feedback control for service, the initialization procedure can be cumbersome too. The output from the feedback also needs to be initialized to insure bumpless transfer. There are several means which can be employed for initialization, a simple method is listed as follows:

Let u(0 − 1) equal the setpoint feedback from inner loop or position feedback from valve position.

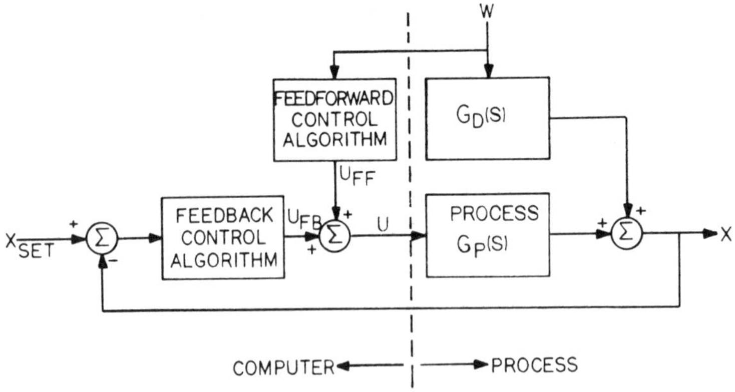

FIGURE 7.12 The block diagram for a combined simple feedback and feedforward control structure is illustrated. The control output, u, is generated from the combination of the output from a feedback algorithm and a feedforward algorithm.

$$u_{ff}(0) = -\left(\frac{k_d}{k_p}\right)\left(\frac{T_p}{T_d}\right) w(0) \tag{7.27}$$

$$u_{fb}(0) = u(0 - 1) - u_{ff}(0) \tag{7.28}$$

Thus,

$$u(0) = u_{fb}(0) + u_{ff}(0) = u(0 - 1) \tag{7.29}$$

As can be seen from this example, the initialization procedure for position form of the feedback and feedforward combination requires several calculation steps. However, the incremental algorithm requires a much simpler initialization procedure just like that for the PID feedback algorithm. The incremental form of the lead-lag algorithm can be written as follows,

$$\Delta u_{ff}(n) = -\frac{k_d}{k_p}\left[\Delta w(n) + \frac{T_p - T_d}{T_d}(\Delta w(n) - y(n))\right] \tag{7.30}$$

where y is intermediate variable

$$y(n + 1) = y(n) + \frac{T}{T_d}(\Delta w(n) - y(n)) \tag{7.31}$$

and

$$\Delta w(n) = w(n) - w(n - 1) \tag{7.32}$$

Suppose the incremental output from the feedback control algorithm is Δu_{fb}, the combined output from feedforward and feedback control algorithm is

$$u(n) = u(n - 1) + \Delta u_{fb}(n) + \Delta u_{ff}(n) \tag{7.33}$$

Comparing Eqs. (7.26) and (7.33), we can see that the computation procedures are different, but the final values are the same.

For the velocity form, the initialization required for the bumpless transfer is set in the following standard manner

$$\Delta u_{fb}(0) = 0 \tag{7.34}$$

$$\Delta u_{ff}(0) = 0 \tag{7.35}$$

As discussed in the last section, the integration of the incremental outputs can be implemented as part of the inner loop, thus, eliminating the need for transferring $u(0 - 1)$ to the outer loop.

One important advantage for using the incremental algorithm can be demonstrated from this example, that is the minimization of process upsets due to equipment failures. When either the feedforward signal or the feedback signal fails, or even when both of them fail, due to any reason, to such a sensor failure, the combined signal $u(n)$ will not be disturbed much because the magnitudes of $\Delta u_{ff}(n)$ and $\Delta u_{fb}(n)$ are usually set to zero. Thus, $u(n) = u(n - 1)$. Operators can be alarmed, and manual actions can be taken to correct the problem without any perturbation to the process.

7.4 THE DERIVATION AND TUNING OF FEEDBACK CONTROL ALGORITHMS

The block diagram of a simple feedback control loop shown in Fig. 7.13 is selected here for discussion. The process transfer function, $G_p(s)$, may include a control valve, process itself, and sensor dynamics. The controller transfer function, $G_c(s)$, is defined to be the ratio of the command output from the controller to the error signal which is generated from the difference between the setpoint and the measured process variable.

The objectives of the control algorithm design are to find the transfer function, $G_c(s)$, such that the specified control performance criteria are satisfied. A simple derivation is given below:

The error signal $E(s)$ is

$$E(s) = R(s) - X(s) \tag{7.36}$$

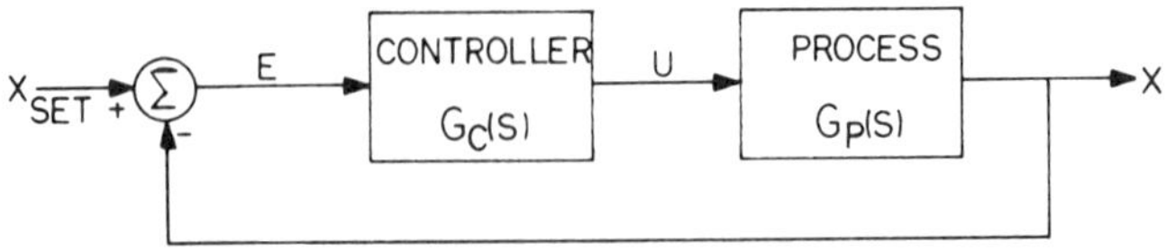

FIGURE 7.13 A typical block diagram of a simple feedback control system is illustrated. The process output is designated with, x. The setpoint of the control system is designated with, x_{set}. The error signal is E and the controller output is u.

The process output X(s) is

$$X(s) = G_p(s)U(s) \tag{7.37}$$

The controller output U(s) is

$$U(s) = G_c(s)E(s) \tag{7.38}$$

By substituting Eq. (7.37) and (7.38) into Eq. (7.36), the controller transfer function becomes

$$G_c(s) = \frac{1}{G_p(s)} \frac{\dfrac{X(s)}{R(s)}}{1 - \dfrac{X(s)}{R(s)}} \tag{7.39}$$

By specifying a realizable closed-loop transfer function, X(s)/R(s), for a given process transfer function, $G_p(s)$, the controller function shown in Eq. (7.39) can be defined. Suppose the transfer function of the process, $G_p(s)$, is given by $k_p/(\tau s + 1)$ which is a first order process. The control performance criterion is assumed to be that no steady state error will result from a step change in the setpoint. The closed-loop response to the step input in setpoint is assumed to be a first order type.

This is shown in Fig. 7.14. For this criterion, the closed loop transfer function can be selected as follows

$$\frac{X(s)}{R(s)} = \frac{1}{\tau_c s + 1} \tag{7.40}$$

Substituting Eq. (7.40) into Eq. (7.39), the controller transfer function $G_c(s)$ is obtained.

$$G_c(s) = \frac{\tau_p}{\tau_c k_p}\left(1 + \frac{1}{\tau_p s}\right) \tag{7.41}$$

This is the transfer function for a proportional and integral control algorithm. The discrete form of the control algorithm for Eq. (7.41) becomes

$$\Delta u(n) = u(n) - u(n-1) = k_1 (E(n) - E(n-1) + k_2 E(n)) \tag{7.42}$$

where

$$k_1 = \frac{e^{-T/\tau_p}(1 - e^{-T/\tau_c})}{k_p(1 - e^{-T/\tau_p})} \tag{7.43}$$

$$k_2 = (1 - e^{-T/\tau_p})/e^{-T/\tau_p} \tag{7.44}$$

and

T = Sampling period

τ_p = Process time constant

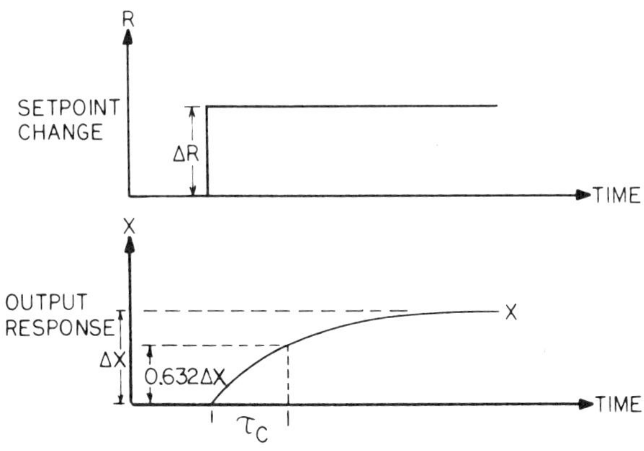

FIGURE 7.14 A graphic representation of the first order response to step input is indicated, where τ_c is the time constant of the first order process.

τ_c = Time constant for closed loop transfer function

k_p = Process gain

When the sampling period T is much shorter than the τ_c and τ_p, that is

$$T/\tau_c << 1$$

$$T/\tau_p << 1$$

then the following exponential terms can be approximated by neglecting the higher order terms. That is

$$e^{-T/\tau_c} \simeq 1 - T/\tau_c \qquad (7.45)$$

$$e^{-T/\tau_p} \simeq 1 - T/\tau_p \qquad (7.46)$$

Substituting Eqs. (7.45) and (7.46) into Eqs. (7.43) and (7.44), the proportional gain, k_1, and integral constant, k_2, become the following simple forms

$$k_1 = \frac{1 - T/\tau_p}{k_p \tau_c/\tau_p} \approx \frac{1}{k_p \tau_c/\tau_p} \qquad (7.47)$$

$$k_2 = \frac{T/\tau_p}{1 - T/\tau_p} \approx \frac{T/\tau_p}{1} \qquad (7.48)$$

The above calculations show that to control a first order process, a proportional plus integral control algorithm can be used. The tuning parameters for this control algorithm can be determined mathematically, based on the process model parameters and the desired performance.

To relate the process parameters, such as process gain or time constant, and control performance parameters, such as τ_c in this case, to the controller tuning parameters, such as k_1 and k_2, is extremely important for adaptive control (8–11). In a simple adaptive control, the process parameters identified on line in real time can be used for k_1 and k_2 updates, after the decision making process in the control decision is completed (12).

7.5 APPLICATION OF LINEAR CONTROL DESIGN TO NONLINEAR PROCESSES

Most control algorithms are based on linear control theory. They can be widely used because most processes can be considered as linear within a small operating range. This explains why many controllers can only be used after the process is started manually and charged up to a certain operating level. Linear control algorithms such as a PID may work well for some processes and not so well for others. This section is written to discuss mathematically how a linear controller works for a nonlinear process.

As discussed in the previous chapter, the Taylor series approximation for a nonlinear process $dx/dt = f(x,u)$ at the neighborhood of operating level (x_s, u_s) is

$$\frac{d\Delta x}{dt} = \frac{\partial f(x_s, u_s)}{\partial x} \Delta x + \frac{\partial f(x_s, u_s)}{\partial x} \Delta u + \text{H.O.} \tag{7.49}$$

where

x = Controlled variable (process output)

u = Manipulated variable

H.O. = High order terms

and

$$\frac{dx_s}{dt} = f(x_s, u_s)$$

$$\Delta x = x - x_s$$

$$\Delta u = u - u_s$$

Now let

$$A = \frac{\partial f(x_s, u_s)}{\partial x} \tag{7.50}$$

$$B = \frac{\partial f(x_s, u_s)}{\partial u} \tag{7.51}$$

Equation (7.49) can be reduced to

$$\frac{d\Delta x}{dt} = A\Delta x + B\Delta u + \text{H.O.} \tag{7.52}$$

Figure 7.15 shows a schematic representation of the nonlinear process and how it is related with a feedback controller with position output u. Based on the previous discussions on incremental models and control algorithms, the simplified block diagram for the control loop is shown in Fig. 7.16.

The nonlinear process is now represented by a linear model with higher order terms and two bias inputs, u_s and x_s. The higher order terms H.O. may or may not be negligible. The process parameters A and B are extremely important for the control algorithm designs which are usually the function of the current operating level (x_s, u_s) and process parameters, such as process gain or process time constant.

The tuning parameters which are utilized in the control algorithm can be adjusted according to the process parameters. This is a type of adaptive control and will be discussed later.

For most of the processes, steady state solutions exist and are constants, therefore,

$$u_s \rightarrow u_{ss} \tag{7.53}$$

$$x_s \rightarrow x_{ss} \tag{7.54}$$

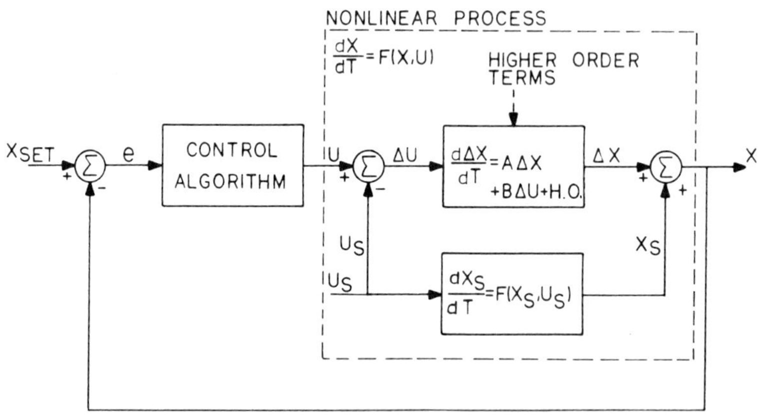

FIGURE 7.15 A schematic representation of the nonlinear process controlled by a position type feedback algorithm is illustrated. The nonlinear process is approximated mathematically by the combination of a linearized model and a steady state model (or an operating model at a given operating level). This is to show the applicability of linear control to nonlinear processes.

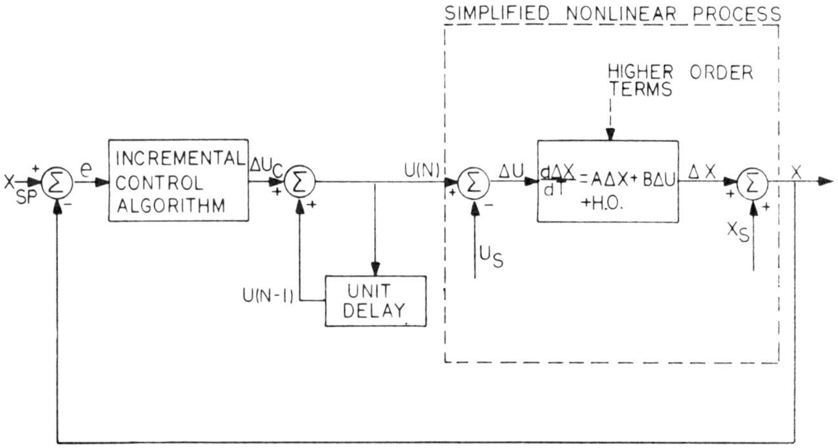

FIGURE 7.16 By observing Fig. 7.15, a nonlinear process can be approximated with a linear process and two additive bias signals, u_s and x_s. This simplified nonlinear process can be controlled by the incremental algorithm with integral control for control output. This simplified linear representation for a nonlinear process provides insight to the applicability of the linear control design.

where u_{ss} is a steady state constant input and x_{ss} is a steady state constant output. At steady state

$$\frac{dx_{ss}}{dt} = 0 \tag{7.55}$$

therefore,

$$f(x_{ss}, u_{ss}) = 0 \tag{7.56}$$

where x_{ss} is the steady state output and u_{ss} is the steady state input. For a sampled data system, u(n) and $\Delta u_c(n)$ are held constant during the sampling interval T. However, the process output variable x is not constant during the sampling interval. These are shown in Fig. 7.17. For a nonlinear process, the linearized process output variable x(n) can be computed as follows:

$$x(n) = x_{ss} + \Delta x(n) \tag{7.57}$$

where the incremental output variable $\Delta x(n)$ can be obtained by solving Eq. (7.52). Assume the coefficients A and B shown in Eq.

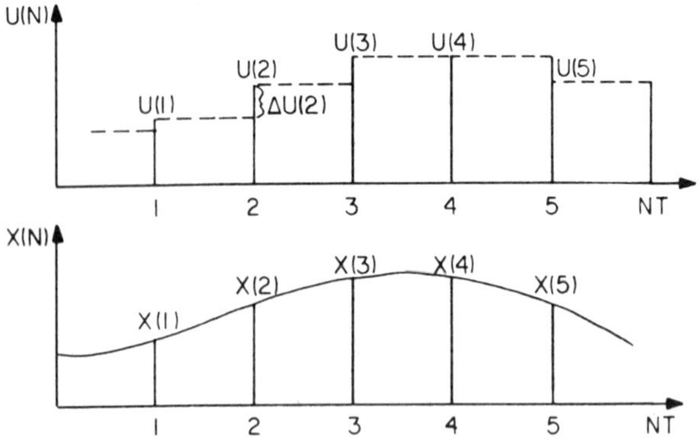

FIGURE 7.17 The sampled input data and the sampled output of the process variables are indicated. The input, u(n), is held constant during the sampling period, T. The process output, x, is not constant during the sampling interval. Usually, u(n) is a supervisory command for the setpoint, which is constant during the sampling interval.

(7.52) are constants for a small range of Δx and Δu with negligible higher order terms. The solution to Eq. (7.52) within the sampling interval T becomes

$$\Delta x(n) = e^{AT}\, \Delta x\,(n-1) + \frac{B}{A}\,(e^{AT} - 1)\, \Delta u\,(n-1) \tag{7.58}$$

Substituting Eq. (7.58) into Eq. (7.57) and rearranging it, we have

$$x(n) = e^{AT}\, x(n-1) + \frac{B}{A}\,(e^{AT} - 1)\left(u(n-1) - u_{ss} - \frac{A}{B}\, x_{ss}\right) \tag{7.59}$$

7.6 MULTIVARIABLE OR COORDINATED CONTROL

7.6.1 Coordinated Control Concepts

The control algorithms we have discussed so far tend to be restrictive to the one input- and one output-type of control strategy. This type of algorithm has been successfully applied in the basic control loops, such as for the flow loop, pressure loop or level loop, etc. These basic control loops are the first level devices for controlling a process as shown in Fig. 7.18. All of the conventional analog control systems and most of the digital distributed control systems are usually designed for controlling processes at this first level. Control panels (or consoles) are used to provide communication between the controllers and plant operators, and operators control the processes by observing several process variables and adjusting setpoints of the first level control loops.

The charactristic of the second level control is more or less a supervisory type. The communication between second level and first level is less frequent, as compared with that between the first level and the process. The strategies used by the operators for adjusting the setpoints of the first level controllers are much more sophisticated than those used to design simple controllers. Operators have to select the proper variables to manipulate, estimate change magnitudes, and change them at the right times. Some of the second level control can be implemented by using the computers. The type of functions are defined as coordinated control. When properly designed, coordinated control will yield substantial control improvement over manually controlled processes.

Coordinated control algorithms have the following characteristics:

1. Multi-input and multi-output control concepts are utilized.
2. Logical functions are involved for decision making.

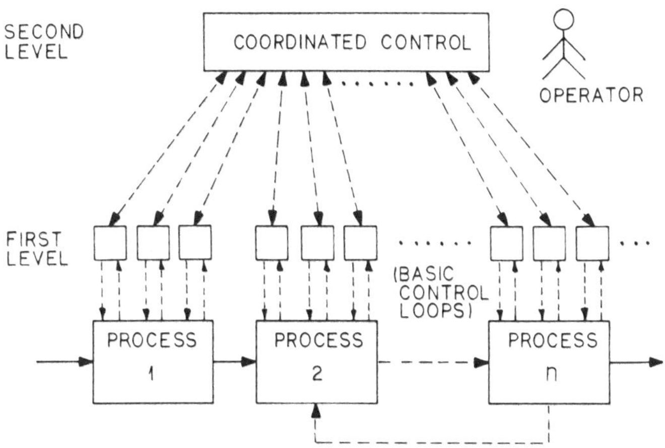

FIGURE 7.18 Control of large processes using a two-level control structure is illustrated. The first-level control is usually performed by the basic controllers. The second-level control is usually performed by operators or computers in a coordinated manner.

3. Global stability of the plant and overall plant control performance are emphasized.
4. Control commands must be issued with correct time and magnitude.

Figure 7.19 shows a block diagram of a coordinated control system structure.

7.6.2 Examples of Coordinated Control Algorithms

There are several algorithms which can be applied for controlling continuous processes or batch processes. These are listed here for discussion.

7.6.2.1 Extended PID Algorithm

This algorithm is implemented as shown in Fig. 7.20. Basically, it is a combination of feedback and feedforward with two switching parameters introduced. Therefore, this algorithm can be used to perform any of the following: feedback PID control only, open loop feedforward, bang-bang control, combined feedback and feedforward control, and combined feedback and bang-bang control.

This extended algorithm is expressed as follows:

$$u(n) = \text{Sat}\ [u(n - 1) + \Delta u] \tag{7.60}$$

where

$$\Delta u = K_{s1}\ \Delta u_{fb} + K_{s2}\ \Delta u_{ff} \tag{7.61}$$

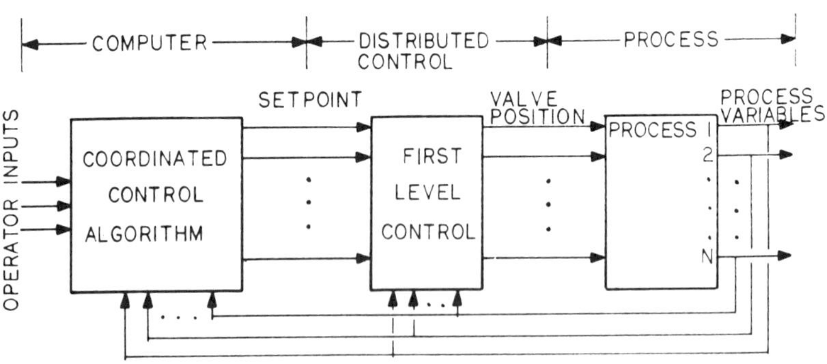

FIGURE 7.19 Coordinated control structure is translated into basic feedback control loops implemented by the distributed controllers and coordinated control algorithms implemented by the computers to segregate the control responsibilities.

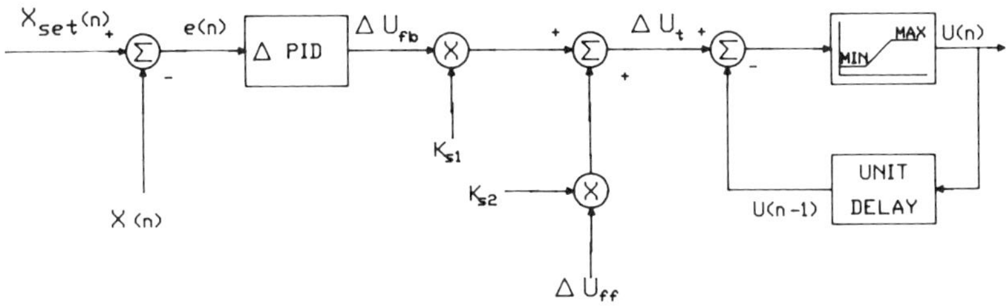

FIGURE 7.20 An extended PID algorithm is a feedback and feedforward control algorithm expressed in a discrete velocity form. The output of the total incremental value, Δu, is integrated with anti-windup nonlinear saturation imposed. Two switching parameters, K_{s1} and K_{s2}, are introduced for superior command by coordinated control activities. Therefore, any combination of control mode can be introduced during the entire control period.

$$\Delta u_{fb} = K_p[K(e(n) - e(n-1)) + K_I\, e(n) + K_D\,(e(n) - 2\,e(n-1) + e(n-2))] \qquad (7.62)$$

and

Δu_{ffi} = Incremental value from feedforward controls

$= \sum_{i=1}^{N} \Delta_{ffi}$

K_{s1} = Superior switching command for feedback control, either 1 or 0

K_{s2} = Superior switching command for feedforward control, either 1 or 0

K = Proportional mode switching parameters, either 1 or 0

K_p, K_I, K_D = PID tuning constants for three mode control

$\text{Sat}[x] = x$, if $x_{min} < x < x_{max}$

$= x_{min}$, if $x \leqslant x_{min}$

$= x_{max}$, if $x \geqslant x_{max}$

Δ_{ffi} = i th component of incremental feedforward control

Note that the switching control parameters K_{s1} and K_{s2} can be used to implement open- or closed-loop control strategies based on process conditions. During normal operating conditions, both K_{s1} and K_{s2} should be set to 1; thus, both coordinated control actions and feedback control actions are combined to regulate the process. During start-up or shutdown periods, or when the process is under the influence of major outside disturbances, such as during a change in crude type for a crude unit, K_{s1} can be set to zero and K_{s2} can be set to 1 so an open-loop control strategy can be used.

7.6.2.2 Single-Input—Multi-Output Algorithm

There are many industrial processes where the process output variable, x, is regulated by manipulating several process input variables, u_j (j = 1, 2,. . . n). The process input/output is shown in Fig. 7.21. Three examples are given here to illustrate the practical applications.

Example 1. As indicated in Fig. 7.9, both oil and gas are used as fuels to heat the feed in the furnace. In this case, the manipulatable input, u_1, is fuel gas flow rate and u_2 is fuel oil flow rate. Both fuels can be adjusted to control the process outlet temperature. This is shown in Fig. 7.22.

Example 2. As indicated in Fig. 7.23, the total feedflow is split into several parallel passes, each pass flow is maintained by a dedicated controller. The total flow rate, x, is measured upstream of these controllers.

The production level of the process is regulated by the total flow rate, x, which can be regulated by adjusting the individual flow controller for each pass. It is very important to point out here that the sum of each pass measured flow rate is not necessarily equal

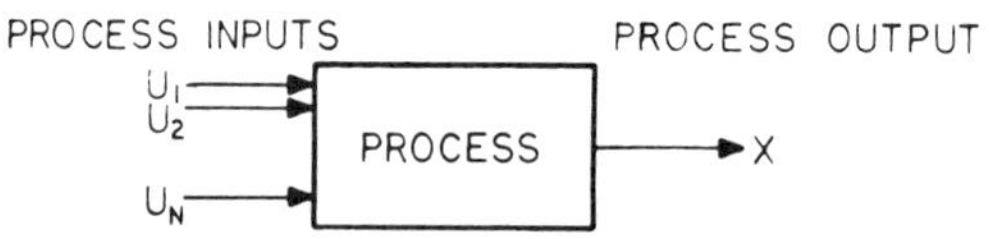

FIGURE 7.21 Multi-input/single-output processes are found in many industrial processes. Control of such processes using a coordinated control concept is very important to provide the interface techniques for control and optimization.

The process output, x, can be regulated to the setpoint of x with u_1, u_2. . . u_n manipulated. At the same time, the control profile, u_1, u_2. . . u_n, can be reshaped to arrive at the optimum condition defined by other criteria.

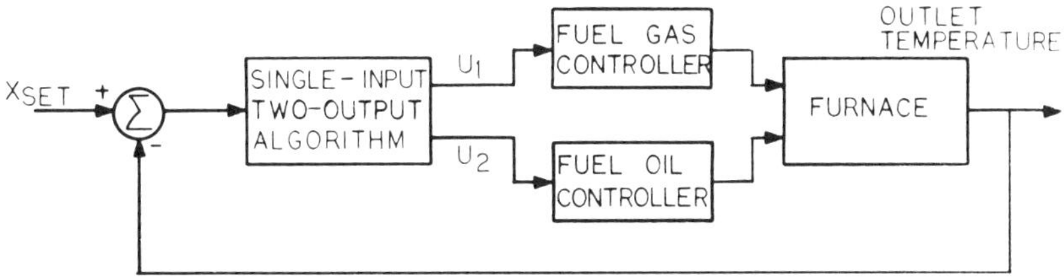

FIGURE 7.22 Furnace outlet temperature control is a typical example of a multi-input/single-output process, which can be regulated by a coordinated control algorithm using a one-input/two-output algorithm. The optimum profile, u_1, u_2 (or ratio), can be determined from economical and operational criteria. Therefore, optimization and regulation purposes are achieved.

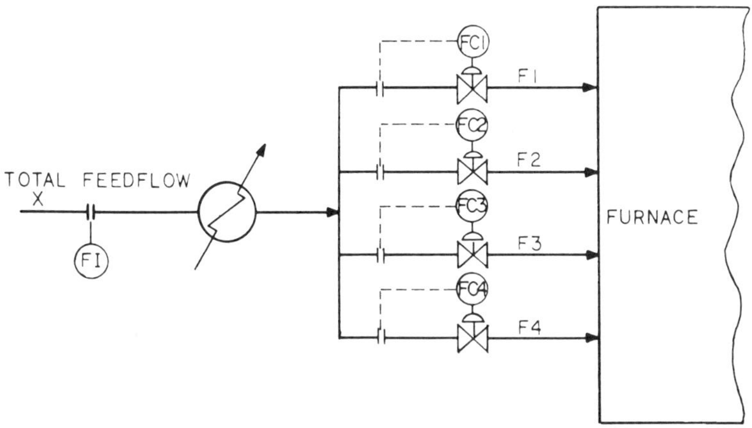

FIGURE 7.23 Total feedflow control, with multipasses controlled individually, is an example of a multi-input/one-output process. The total flow, x, can be regulated to the setpoint value, x, by manipulating FC_1, FC_2, FC_3, and FC_4. At the same time, there is another operational criteria to be satisfied, such as maintaining the same outlet temperature for each pass. Therefore, the flow profile (relative flow ratio) must be adjusted accordingly. This type of control can be achieved using coordinated control algorithms, such as extended PID.

calibrating errors of transmitters or partial vaporization of feed. The total flow control block diagram can be expressed as shown in Fig. 7.24.

The coordinated command and individual pass flow rate is related as follows:

$$F1_{set}(n) = u_1(n) \tag{7.63}$$

$$F2_{set}(n) = u_2(n) \tag{7.64}$$

$$F3_{set}(n) = u_3(n) \tag{7.65}$$

$$F4_{set}(n) = u_4(n) \tag{7.66}$$

$$F \text{ total flow} \neq \sum_{i=1}^{4} Fi \text{ measured} \tag{7.67}$$

Usually $u_1 \neq u_2 \neq u_3 \neq u_4$. The exact control profile $u_i(n)$ is determined by other criteria, such as the balance of furnace exit temperature of all passes.

Example 3. A dryer with four hot air circulation chambers is utilized to control the product moisture by manipulating the temperature profile. This is shown in Fig. 7.25. The moisture content of

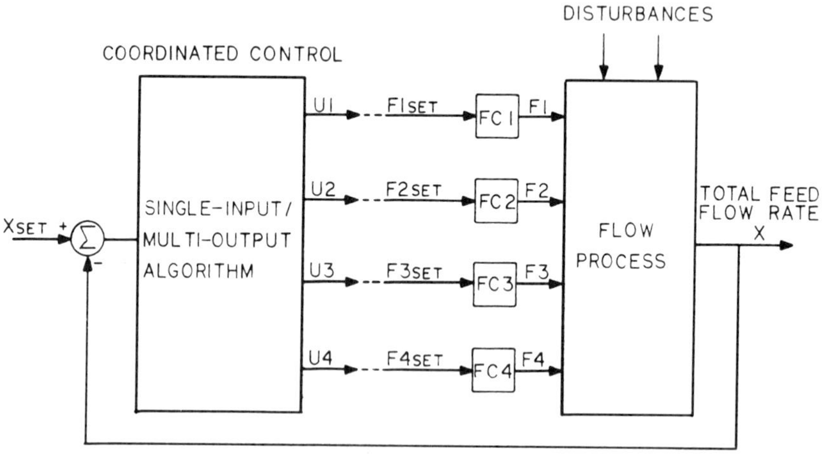

FIGURE 7.24 Total feedflow control implemented with a single-input/multi-output algorithm is arranged in a coordinated control structure as shown in Fig. 7.19. This is also an example of a multi-input/single-output process.

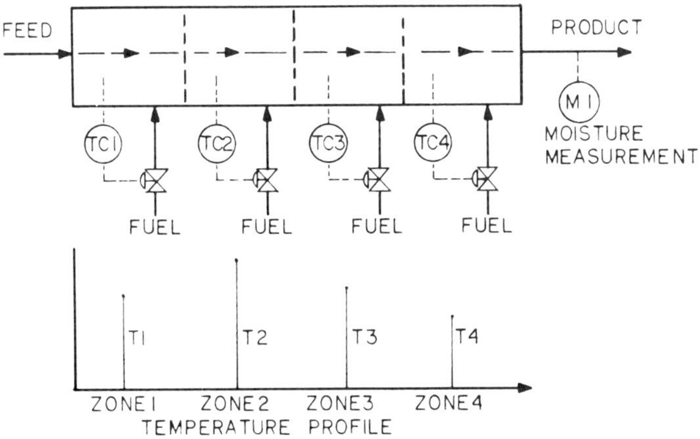

FIGURE 7.25 Moisture control of the drying process using temperature control at each zone to manipulate the input profile. This is also an example of a multi-input/single-output process.

the product can be regulated by adjusting the four temperature zones. The arrangement of the control scheme is similar to the total flow control scheme shown in Fig. 7.24, except the flow controllers are replaced by the temperature controllers in this case.

A generalized implementation diagram is illustrated in Fig. 7.26 for single-input and multi-output coordinated control algorithms. This diagram can be applied to all of those examples discussed above. Note that incremental PID algorithms are used to perform the feedback control. This is desirable for the purpose of bumpless transfer from auto to cascade mode.

7.6.2.3 Ramping Algorithm

This algorithm is most often applied in the start-up or shutdown operation or for large production rate changes. Figure 7.27 shows the ramping command generated from the ramping algorithm. Usually the maximum rate of change increment, Δ, is specified by the operator or engineer in the production facility. The purpose of the ramping command is to increase or decrease the setpoint gradually to prevent a large upset to the process.

The ramping algorithm is defined as follows:

$$u(n) = \text{Sat}_2\,[u(n-1) + \text{Sat}_1\,(u_{set}(n) - u(n-1))] \qquad (7.68)$$

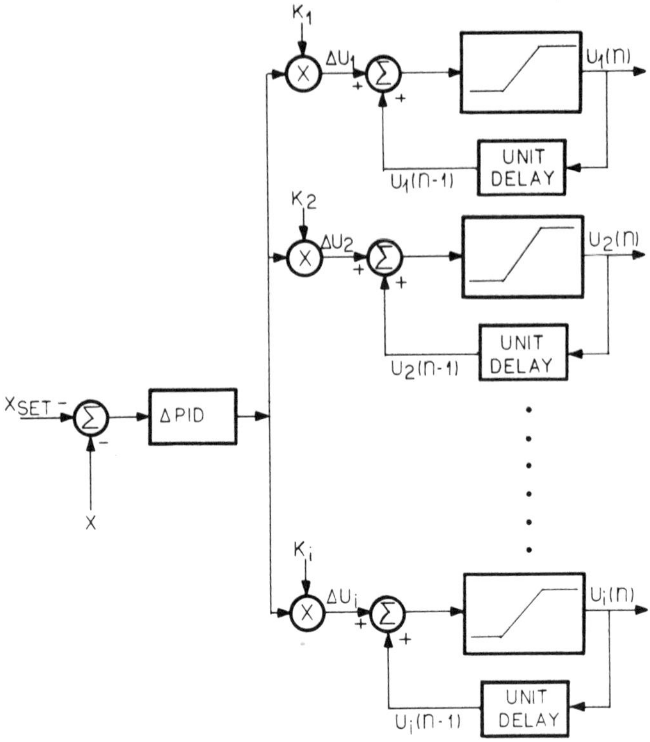

FIGURE 7.26 A schematic block diagram for a single-input/multi-output algorithm is illustrated. This algorithm is useful for controlling a multi-input/single-output process, especially $u_1 \neq u_2 \neq u_3 \ldots \neq u_n$. Usually, an extended PID can be applied to construct a coordinated control.

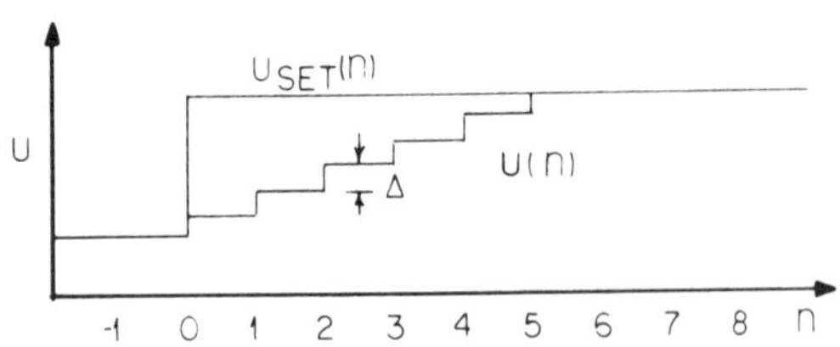

FIGURE 7.27 Ramping command, u(n), is generated from the step command, $u_{set}(n)$, with a fixed increment Δ. The value of Δ is specified by the operators or engineers from the console.

where

$$\begin{aligned} Sat_1(y) &= \Delta \text{ if } y > \Delta \\ &= -\Delta \text{ if } y < -\Delta \\ &= y \text{ if } -\Delta < y < \Delta \\ Sat_2(y) &= u_{max} \text{ if } y > u_{max} \\ &= u_{min} \text{ if } y < u_{min} \\ &= y \text{ if } u_{min} < y < u_{max} \end{aligned}$$

The implementation of this algorithm is shown in Fig. 7.28.

It is seen that the ramping algorithm is generated with two nonlinear blocks, Sat_1 and Sat_2, connected in cascade with a unit delay block connected in a feedback manner. The difference in the saturation nonlinear blocks Sat_1 and Sat_2 is the input parameters only; the computation algorithm is the same. If a generalized saturation algorithm, GSAT, is created as a function block or routine, such as GSAT (X_{max}, X_{min}, X), the ramping algorithm can be expressed as follows

$$\Delta u = GSAT\ (\Delta, -\Delta, u_{set} - u\ (n-1))$$

$$u = GSAT\ (u_{max}, u_{min}, u\ (n-1) + \Delta u)$$

For example, when the total feed rate to a crude unit of a petroleum refinery is to be increased or decreased, a ramping algorithm can be used. In order to maintain a stable operation during the transition period, the total feed rate must be changed gradually to maintain the product quality.

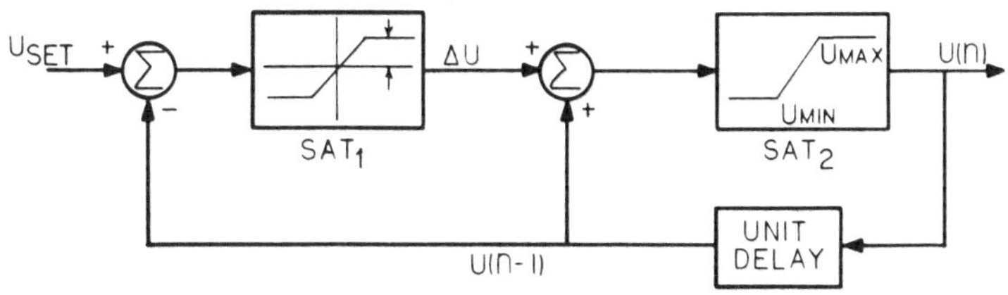

FIGURE 7.28 A ramping algorithm is implemented with two saturation nonlinearities in cascade with the unit delay of output as a feedback. This is implemented as a feedback control type.

7.6.2.4 Algorithm with Switching Logic

In an explosive process, such as combustion, override control sometimes is necessary due to safety reasons. The extended PID algorithm in Sec. 7.6.2.1 can be used for this purpose.

There are several switching parameters built in the extended PID algorithm. Parameter K_{s1} in Eq. (7.61) can be used to turn the closed-loop feedback control on or off without perturbing the process during transition. Parameter K_{s2} also can be used to turn the feedforward control on or off without perturbing the process. This is accomplished by the standardized initialization procedure for velocity algorithm.

The process shown in Fig. 7.25 can be used to illustrate the importance of switching logic in the actual operation. Based on the operator's experiences, there are disturbances applied to the process in a very short time period. One of such disturbances, for example, is the feed on the conveyer belt becoming very wet just for a few seconds and then returning to normal. The furnace temperature operation cannot be changed at such a fast rate to control the product moisture, and a closed-loop feedback control during this short period causes oscillations. In this case, it is necessary to apply a switching logic to change the control mode from a closed-loop feedback control to an open loop control for a short period of time and then back to the closed loop control again when such disturbances have disappeared. The purpose of using the switching logic in conjunction with the switching parameters built in the algorithm is to improve the control performance which most of the linear control approaches cannot accomplish.

For decision making, it is beneficial to collect a substantial amount of information from process input and output for judgement. Therefore, the sampling frequency for disturbance monitoring or process output monitoring should be substantially faster than the frequency of the control command cycle. The control commands are generated after the decision making algorithm has completed its decision process.

Unlike the regular feedback control with a fixed control algorithm, the coordinated control algorithm, with switching logic and logic parameters imbedded in one single equation, has a flexible structure.

Thus, the algorithm can be made more adaptable to the needs of the new process environment whenever the decision logic determines that it exists.

Many adaptive control applications are based on changing the tuning parameters, but the control algorithm structures stay the same. In coordinated control, the control algorithm structure can be changed for each new situation.

REFERENCES

1. Ragazzini, John R. and Franklin, Gene F., *Sampled-Data Control System*, McGraw-Hill, New York, 1958.

2. Ogata, Katsuhiko, *Modern Control Engineering*, Prentice-Hall, Inc., Englewood Cliffs, New Jersey, 1970.

3. Kuo, Benjamin C., *Digital Control Systems*, SRL Publishing Champaign, Illinois, 1977.

4. Tucker, G. K. and Wills, D. M., *A Simplified Technique of Control System Engineering*, Honeywell, Inc., Industrial Division, Fort Washington, Pennsylvania, 1958.

5. Tou, J. T., *Modern Control Theory*, McGraw-Hill, New York, 1964.

6. Smith, Cecil L., *Digital Computer Process Control*, Intext Educational Publishers, San Francisco, 1972.

7. Takahashi, Yasundo, Rabins, Michael J., and Auslander, David M., *Control and Dynamic Systems*, Addison-Wesley Publishing, Reading, Massachusetts, 1972.

8. Astrom, K. J., Theory and Applications of Adaptive Control—A Survey, *Automatica, 19*(5):471–486 (1983).

9. Landau, Yoan D., *Adaptive Control, The Model Reference Approach*, Marcel Dekker, Inc., New York, 1979.

10. Kays, A. and Scheib, T. J., A self-tuning method for Smith predictor and PID controls, Advances in Instrumentation Volume 39, Part 2, Proceedings of the ISA International Conference and Exhibit, Houston, Texas, 1984, pp. 843–849.

11. Dahlin, E. B., Designing and tuning digital controllers Part 1, *Instrument and Control Systems, 41*:77–83 (June, 1968).

12. Lin, C. S., Tsai, T. H., and Lane, J. W., Coordinated control and its implementation, 2nd World Congress of Chemical Engineering, Montreal, Canada, 1981, pp. 423–427.

8

Process Control for Batch Processes

8.1 INTRODUCTION

Chemical batch processes have been generally considered "less than desirable" by both design and operation engineers when compared with continuous processes. When a chemical formulation is first developed in the laboratory, it is usually developed using a batch process. After the formulation is proven, many of the early steps towards commercial scale production are still based upon batch-type pilot plants. But when a high capacity plant is needed, the emphasis is to convert the batch plant to a continuous plant. This requires a considerable research and development effort, and the results have been that the majority of chemical processes are continuous.

However, continuous plants are not suitable for some types of chemical processes. For example, some polymerization and fermentation reactions require very long reaction times and the addition of new chemicals at different stages of operation. A process utilizing a batch reactor is much easier to design than a continuous process to handle this type of reaction. In addition, a batch plant is more flexible than a continuous plant. When a plant is required to produce many different grades of products using the same equipment, it becomes more economical to use batch processes.

The most important objections to batch plants is that batch plants have to be cycled frequently and monitored carefully, thus, requiring higher labor costs per unit of product throughput than continuous processes. Batch processes are seldom at steady state. Furthermore, product quality variation from batch to batch is symptomatic of the lack of precise measurement and control in raw material charge quantities and operating conditions. Note that for a continuous process, operators can spend a long time, usually days, to bring a process to a designed operating level, or to a steady state. Operators can make frequent adjustments in process conditions according to laboratory test results and other off-line measurements. The off-spec materials produced during the start-up transient period can be either reprocessed or blended with the good products during normal operation. Operators for a batch process do not have this luxury. The process is in transition most of the time, and the limited time cycle of a batch does not allow for many corrective actions. Variable product quality and low throughput have always been associated with batch processes.

Modern control systems and computers have definitely brought new life to batch processes. As Nakagawa (1) explained in his 1970 paper, computer technology, when properly applied, can eliminate many of the shortcomings associated with batch processes. Indeed, many high capacity PVC plants, using a batch process, have been built recently with excellent results in consistent high quality products and low direct labor costs. Computer technology has made many major advancements since 1970, and there have been significant increases in the availability of batch control systems being offered by instrumentation and control companies.

Unfortunately, many of these batch control systems are not friendly for inexperienced users, and their potential is not being fully utilized because of the lack of application engineering know-how. This chapter was developed to help all engineers who are involved in the design and operation of batch process plants.

8.2 THE OPERATION OF BATCH PROCESSES

8.2.1 Sequential Operations

The operation of a batch process is time-variant in nature. Practically all batch processes require the following basic operations for one or more reactors:

Preparation
Charge
Reaction
Discharge
Postpreparation

Many batch processes involve variations of the above operations in that the charging operation may continue, in part, simultaneously with the reaction operations, and the discharge operation may be initiated prior to the completion of the reaction operation.

The batch cycle time, i.e., the elapsed time from the beginning of one batch to the beginning of the subsequent batch, is usually measured in hours rather than days, typically greater than an hour, but less than 48 hours.

Figure 8.1 shows a typical batch process plant consisting of a common charge area and multiple reactors. There are also multiple intermediate rundown and final product storage tanks which can be used to store different types of products. To appreciate the control problems for a batch plant, a discussion in the functional requirements for each operation will help. The following discussion is based upon the process configuration shown in Fig. 8.1 as an example.

8.2.1.1 Preparation

The four raw material charge tanks are required to be maintained at reasonable levels. There are several important considerations which greatly affect the results of the down-stream operations:

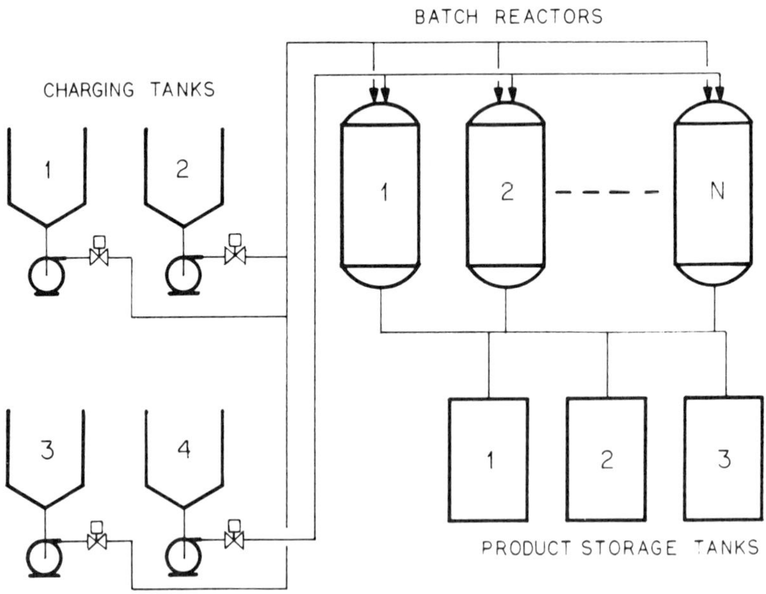

FIGURE 8.1 Simplified process flow diagram for a typical batch plant showing major equipment.

Solution concentration Suppose a solution is to be prepared by operators mixing several different chemicals; the final compositon becomes a very important factor in the reactor operation. A consistent composition will be most essential; otherwise, a compensation in charging quantity may have to be made according to the measured composition.

Time to refill charge tank One may not want to refill a charge tank while that charge tank is being used for reactor charging. There are two reasons for this concern. First, the charge tank may be used to measure how much solution is charged into a reactor, and second, the composition of the solution in the charge tank may not be exactly the same as that used to refill the charge tank. If these restrictions are indeed observed, then the time to refill the charge tank becomes critical because this tank is used to fill all reactors, so the time available for refilling may be very limited.

8.2.1.2 Charge

The *charge* phase is probably the most complicated operation because it is more likely to be affected by operating errors. The reason is that this phase requires multistep, sequential operation and accurate measurements. Because of the large number of operating devices involved (valves, pumps, etc.), equipment failure during an operation is common, thus, making the monitoring of the charging operation extremely challenging.

The goals for charging operation are straightforward. Each solution should be charged at the desired time, each solution should be charged at the desired quantity, and finally, the correct amount of the reactant is ultimately placed in the reactor.

Figure 8.2 shows a time chart for solution charging based on the process flow diagram shown in Fig. 8.1. Note that the charging sequences for the three solutions are started and completed at different times. The reason is the need to assure that at the end of charging one reactor, the charging pipe is filled with only one solution, that is Solution 1. Should the timing for charging not be precisely controlled, then the charging quantities to all reactors would be mixed up.

Solution 4 charging starts some time after an agitator is started; this means this chemical could be charged to all reactors simultaneously, thus, a measurement device different than that used for the first three solution is required.

From the operation's standpoint, there are several factors which make the charging operation difficult:

Multiple grade When each reactor requires different quantities of several solutions.

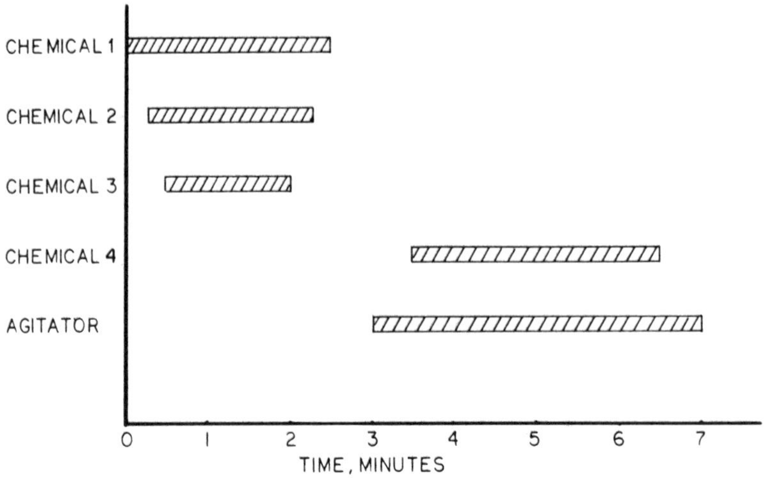

FIGURE 8.2 Time chart for solution charging to a batch plant showing sequenced chemical charging and agitation.

Precise charging quantity Most charging volumes are measured by volumetric flow measuring devices, such as turbine meters, or by flow rate measuring devices such as orifices. For the latter devices, flow integration is also needed to compute total volume. This type of measuring device may not give sufficient accuracy without temperature compensation and/or concentration compensation, and for some plants these compensations must be done manually. Precise charging quantity is a very elusive goal for batch processes requiring manual addition of small quantities of liquid or solid reactants. Even safe operation may not be achievable. Developing methods to add small amounts of liquid or solid reactants to a batch process precisely is a major challenge.

Equipment failure Any of the equipment associated with the charging operation could fail. A valve failure during solution preparation may cause great difficulties for the operators. In addition, a flow meter failure during charging may cause even a greater problem because such failure may not be identified until some time later. To identify equipment failure and to recover from such a failure is probably one of the most difficult, but also desirable tasks for batch process control.

8.2.1.3 Reaction

For most batch chemical reactor operation, the most important function is the control of the reaction conditions. For the majority of batch reactors, the control objective is to maintain the reactor temperature at a desired setpoint. Specifically, a temperature controller which has the following performance characteristics is required.

Permits heating the reactor at the fastest possible rate, but does not permit temperature setpoint overshoot.
Maintains the reactor temperature at the temperature setpoint over the entire batch-cycle within a small tolerance.
Provides consistent performance for all temperature setpoints, that is, there is no steady state offset at temperature setpoints other than the base temperature setpoint.

Since a reactor is usually used to produce different grades of products, each requiring a different set of reaction conditions, the dynamics of the process change from grade to grade. A controller with constant tuning parameters may give good results for one grade, but becomes unstable or sluggish for another grade. To achieve the performance criteria listed above, an adaptive controller is usually needed.

8.2.1.4 Discharge

To prepare a reactor for discharge, an operator must first determine that the reaction is completed, and then to determine that the reactor is safe to discharge, i.e., the reactor has been depressurized and the explosive or toxic chemicals are removed. Another important task for an operator is to discharge the product to the designated post processing equipment and storage tankage. For a plant making a large number of products, the scheduling for reactor discharge could be a major factor.

8.2.2 Recipe

The word "recipe" is used by many plants as a general term which consists of a list of information pertaining to the conditions required to produce a given product, such as the charge rates and total quantities of all solutions, time of charging, reactor temperature setpoint, time for reactor discharge, etc. A plant may have many recipes for a large number of products, and different recipes could be used for different reactors at the same time. To assure the right recipes are used in a reactor, many plants use written instructions for each batch, and require operators to fill out a batch summary report. This increases the operator's routine work load and some times causes confusion. There are substantiated reports that operators have made mistakes in charging quantities or reactor settings, and these resulted in production losses or even accidents.

One should realize that a recipe list could be long. It not only includes the chemical formulation, but also operating conditions for control and monitoring. For example, the time duration for charging a chemical could be a data point on the recipe list, so operators can monitor the charging of that chemical even when the "end of charging" by flow devices has failed. To update many recipes based on new laboratory and operation information, and transmitting them to operators can be a demanding job.

8.2.3 Product Quality and Throughput Optimization

8.2.3.1 Batch Cycle Time Control

A very important word for batch process control is consistency because only when every operation is kept consistent, can the result be reasonably analyzed, and corrective actions or feedback compensations be made. The easiest way to evaluate whether a batch plant is consistent is probably by comparing its cycle time for making identical products. Cycle time can vary greatly for a poorly controlled plant.

The best thing that ever happened to batch plant control is the availability of process control computers, programmable logic controllers, and sophisticated digital controllers. Note that unless a plant is very simple, a good control system must include all three types of these devices. The reason for this requirement will be described elsewhere in this chapter.

To control and optimize batch cycle time, one can consider the implementation of any one of the following functions:

Automatic recipe processing A supervisory control system can be so designed that all recipe information can be automatically transmitted to the proper location after a product code is selected by an operator. For example, the charge rate and total charge quantity can be sent to the flow control devices without operator data entry. This saves time and eliminates mistakes.

Automatic charging control of multiple streams As described before, it may be required that several solutions be charged to a reactor simultaneously. This reduces the total charging time as compared to charging solutions sequentially one at a time. With automatic recipe processing, all charging can be completed without manual intervention.

Advanced reactor temperature control Again, the stated performance criteria for a good temperature controller are: permit the rapid heat up of the reactor and assume control when the batch temperature approaches the desired temperature setpoint without overshooting the temperature setpoint, and subsequently maintain the batch temperature within an acceptable tolerance of the temperature setpoint over the entire batch reaction period, and finally, provide the same good control for other temperature setpoints.

A conventional analog controller has certain functional limitations which can be clearly understood when one examines the mathematical representation of the three-mode algorithm. In any analog controller, the major restriction on the controller performance is that a given set of tuning parameters, that is, gain, integral, and derivative, will result in satisfactory controller performance *only* over a narrow range of process conditions. When an attempt is made to utilize the controller beyond these limits without retuning, the controller performance deteriorates.

For continuous processes where the dynamic characteristics do not change or only change slowly, this limitation on the analog controller is not a hindrance. However, the performance of the analog controller is seriously compromised when used for a batch process where the process dynamics vary over a wide range throughout the batch cycle.

Thus, any system that permits the changing (adaptive tuning) of the controller parameters according to detected changes in process dynamics and product specifications is very desirable. The digital control computer has been used very successfully for this function. A detailed discussion on adaptive control is given elsewhere in this book.

Automatic control of solution preparation Most solution preparation operations mix manual additions of chemicals and charging of solvents or chemicals through flow meters. An automatic system can be designed to give clear instructions to operators on how much and when a chemical should be added, and controls the charging of chemicals that are measured on-line. In addition, the transfer of chemicals from the solution preparation area to the charge tanks can be automated to minimize the waiting time for charge tank refilling.

Automatic control of reactor discharge Two important functions can be achieved. They are identifying that the reactor is ready for discharge and opening the correct valves for the product to go to the correct storage tank. This could save significant time delays for some plants.

Automatic batch report generation Actual operational results can be automatically summarized into one report for future evaluation. Data, such as the actual charging quantity of a chemical (which may or may not be identical to what the recipe calls for), temperature profile, reaction time, alarm conditions, and even laboratory test results of product quality, can be stored in one report. This will provide considerable help for plant troubleshooting and product improvement programs.

8.2.3.2 Automatic Production Scheduling

When there are many grades of products to be made, and each grade has its own recipe, and can require a different cycle time to produce it, the scheduling of reactors may become a complicated task. As described earlier, many batches may share the same set of equipment, such as several reactors sharing the same charging system. Poor production scheduling may cause unnecessary idling of major equipment, such as a reactor, thus, reducing plant throughput.

There are many automatic production scheduling techniques developed and applied by the electronic data processing staffs. Conceptually, a good automatic scheduling system allows a user to enter all existing orders into a computer, and computes a production schedule to minimize lost time and to meet all constraints such as delivery dates.

This type of information can be found in most computer application books and will not be repeated here. There are, however, several important factors one must consider before automatic production scheduling can be effectively applied:

> The batch cycle time for a product should be consistent. For a manually controlled batch process, or a poorly designed automatic controlled process, batch cycle time for a given product is not consistent. A nominal eight-hour cycle time for making a product may mean six to ten hours in actual operation. When this is the case, an automatic production scheduling may not be effective in reducing equipment idling. The reason is obvious; when the actual cycle time deviates too much from the estimated cycle time, scheduling becomes meaningless.
>
> Automatic scheduling is a data processing function and should be implemented by using the data processing computer because this function requires frequent data entry, file management, and communications to other data processing functions. Often an attempt is made to implement this function on a control computer. It can be done, but results in an unnecessary complication of the control computer software.

8.2.3.3 Optimum Reactor Temperature Profile

A discussion was presented in the previous section on the changing of the control algorithm parameters (adaptive tuning) to reflect changes in process dynamics solely for the purpose of improving the controller performance. The changes in process dynamics were caused by both those which were predetermined, such as the reactor temperature setpoint or the load, and those detected throughout the batch cycle, such as the change in the heat transfer coefficient of the reactor wall.

The benefit of this improved controller performance is primarily improved product quality. Increased reactor throughput does not directly result from improved temperature control. However, increased reactor throughput can be achieved as a result of improved reactor control and the use of additional capabilities of the digital controller. Two examples of improved reactor performance, which can result from the advanced control capabilities of digital controllers, will be discussed. In order to present the discussion on a more quantitive basis, an example batch reaction model was utilized. The simple polymerization of vinyl acetate in toluene using the initiator, azo-bis isobutylene (AZO) was used in the model. The constants used in the expression for reaction rate were typical of those given in the literature (2–4). The model calculates incremental reaction rates, product and reactant concentrations, and heat of reaction for small time increments and then prints out the respective profiles.

Input variables, including monomer and initiator concentration, temperature setpoint, and mode were adjustable. The model equations are:

Decomposition of Initiators:

$$DI = 2FK_D I60$$

where

DI = Moles of free radicals formed per minute per liter

F = Efficiency factor assumed to be 1.0

K_D = Rate constant

I = Initiator concentration, moles per liter

$$K_D = 2 \times 10^7 e^{-13.7/0.00108T}$$

T = Temperature in Kelvin

Polymer Formation:

$$R_p = K_p M \, \Sigma DI \; 60$$

R_p = Monomer converted to polymer mole per liter per minute

K_p = Reaction rate constant

M = Monomer concentration moles per liter

ΣDI = Moles of free radicals formed, moles per liter

$$K_p = 524e^{-4.7/0.00198T}$$

Heat of Polymerization:

$$DQPP = 21.3\ R_p,\ K\ cal/min/liter$$

To evaluate the effects of time-temperature profiles on production rates, two modes of operation were simulated by computer simulation. These were: (1) maintain a constant temperature profile over the entire batch, and (2) maintain a constant temperature until the peak conversion rate was reached and then ramp the temperature to maintain nearly constant conversion until the rate cannot be sustained with a modest increase in temperature.

For some products of batch processes, the product quality is much more sensitive to reaction temperature than initiator concentrations. If this is indeed the case, then considerable plant increases can be achieved by maximizing the initiator concentrations (within the constraints of the reactor dynamics constraints) and reducing the batch cycle time.

Figure 8.3 gives a typical reaction rate versus time curve for the process used as an example. The end of a cycle has been defined

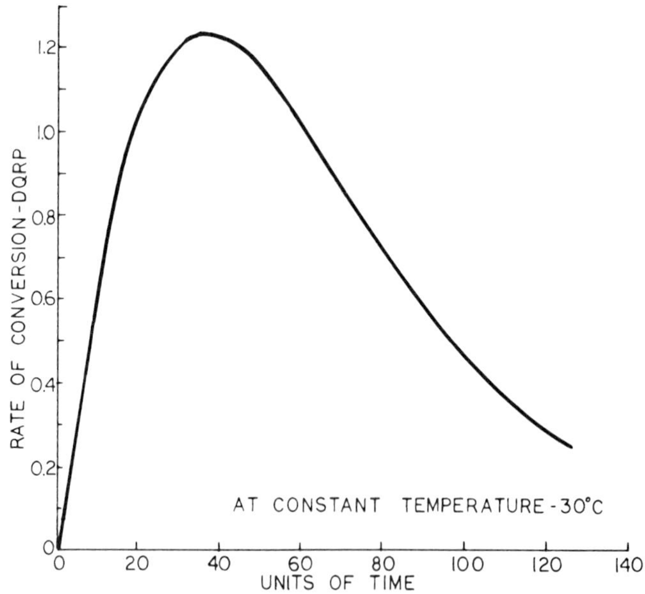

FIGURE 8.3 Typical time profile for exothermic batch chemical process showing rapid increase in rate of conversion followed by a slow decline as the reactant is consumed.

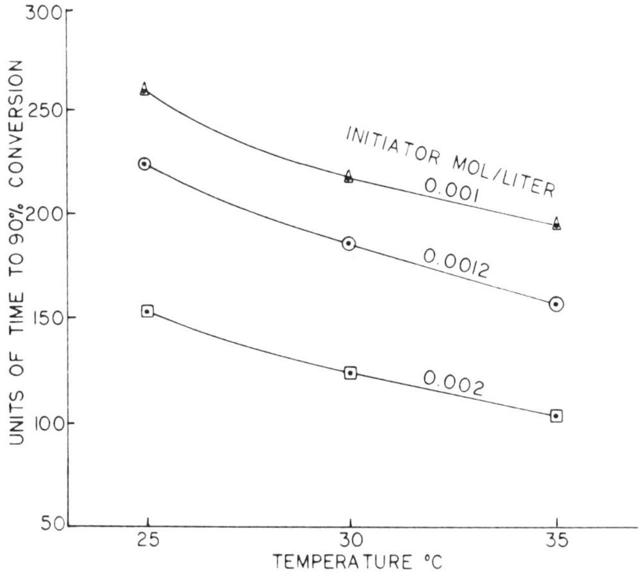

FIGURE 8.4 Effects of initiator concentration and reaction temperature on elapse batch time required to obtain 90 percent conversion. Elapsed time decreases with increased temperature and initiator concentration.

as the time of 90 percent conversion of the monomer. The curve in Fig. 8.3 is the base curve for the process. The temperature was held constant at 30°C, and the initiator concentration was 0.001 moles per liter of charge. As the initiator decomposes, forming new radicals, and monomer concentration is large, the rate of reaction increases. The rate reaches a maximum when the initiator is expended and only the polymerization reaction continues consuming monomer. The reaction rate slowly declines as the monomer is consumed.

Increasing the initiator concentration has two effects, assuming no temperature change. Increasing the initiator results in a higher maximum conversion rate and a shorter time required to reach 90 percent conversion. These two effects are shown in Fig. 8.4 and 8.5.

Now one can take advantage of this process knowledge in control system design. When a reactor is designed and built, the maximum heat transfer coefficient between the batch and the water jacket is determined. However, it does not remain unchanged. The heat transfer coefficient will be less from time to time due to internal fouling in the reactor and water jacket fouling. But even though the heat transfer coefficient does change, the present value at any given time can be calculated *if* accurate process information is available. In most cases having an accurate measurement of the jacket water inlet and outlet temperatures and the jacket water flow will be sufficient to calculate the effective heat transfer coefficient. Also the maximum rate of heat generation can be calculated for each batch to provide feedback information to the recipe model.

Assume that for a reactor operating a 30°C the maximum rate of heat reaction which could be controlled is 0.9 Kcal per liter, then the initiator concentration in the base case, which was 0.001 mole per liter, could be increased to about 0.00123 mole per liter. This would have the effect of reducing the batch cycle time from 218 minutes to about 175 minutes. The increase in plant capacity would be significant and the investment required to accomplish this function would be minimal.

The results of many theoretical studies indicate that an optimum time-temperature profile exists for many batch reactions that will maximize product yield, minimize batch cycle time and/or maximize product throughput (5).

This can be demonstrated by referring to the example again. Assume that the initiator concentration is at maximum for the heat transfer coefficient, and that the product quality is not sensitive to temperature variations after the initiator is depleted.

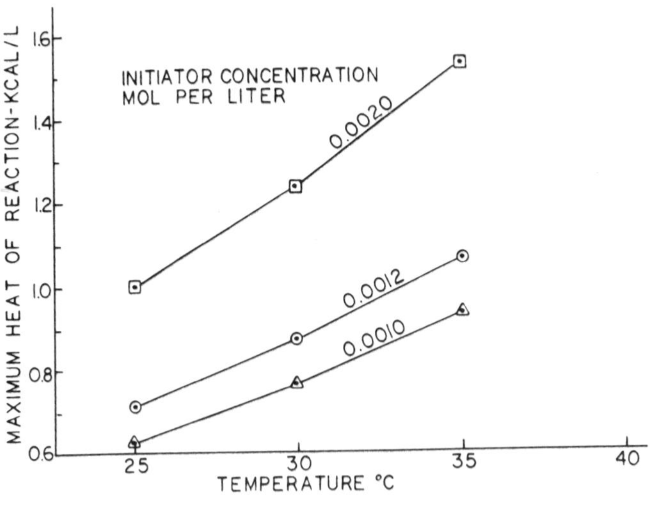

FIGURE 8.5 Effects of initiator concentration and reaction temperature on maximum best of reaction. Maximum best of reaction increases with increasing initiator concentration and reaction temperature.

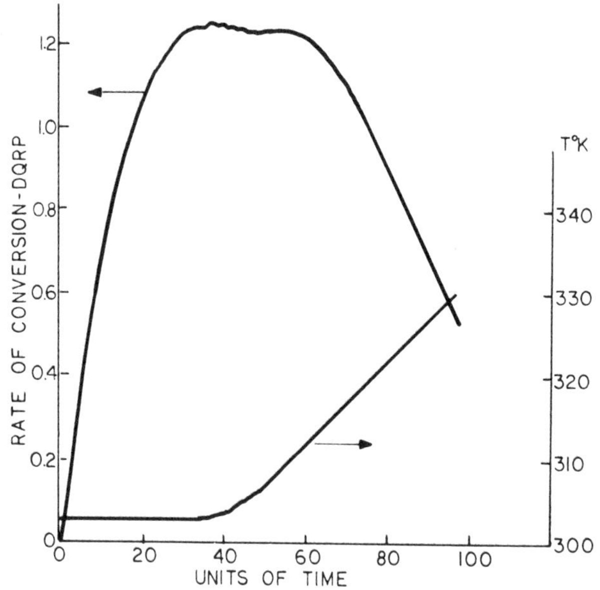

FIGURE 8.6 Effect of ramping batch reaction temperature point where maximum rate of conversion occurs; thus, it decrease batch time required to reach 90 percent conversion.

The profile in Fig. 8.6 was generated by first holding the temperature setpoint constant until the maximum heat of reaction was reached, then increasing the temperature a half a degree if the heat of reaction for the present time increment was less than that for previous time increments. The resulting profile shows the temperature being held at 303 K to time increment 35, slowly decreasing to time increment 49 and then increasing at a rate of one-half a degree per minute to 330 K at the time increment 97, at which time the batch was completed. The 90 percent conversion was reached 28 minutes sooner than the base case. Again, this would result in a significant increase in plant throughput capacity.

8.3 CONCEPTS FOR BATCH PROCESS CONTROL SYSTEM DESIGN

8.3.1 Conventional System

The conventional batch process control system generally consists of simple analog instruments and limited application of automatic sequence control, usually in the form of cam controllers and relay interlocked logic. Operators have performed most of the logic checks and sequence actions manually either by directly operating the valves and motors, or by push buttons located on the control panel and remote valve operators and motor starters. All reactor header valves are manually opened and closed by the operators. A flow integrator is used for charging solutions, but the temperature compensation is calculated by the operator, and the charge pump is manually shut at the desired volume. The reactor heat-up and temperature control are accomplished by the same analog temperature controllers.

Instrument engineers expended a great amount of effort trying to design automatic sequence control devices long before solid-state logic devices and digital computers were available. Dr. Amrehn (6) reported that Hules of Germany designed an automatic batch control system in 1954. This system employed relay logic, flow meters, pulse counters, limit switches, analog controllers, etc., for sequencing and reactor control of a polyvinyl chloride plant. Most of the system was removed shortly after plant start-up because it was plagued with unceasing hardware failures.

There are many reasons for the failure of automatic batch control systems in the early days of automation. The most important reasons are probably the limited capacity of conventional devices which make it impractical to design a complete automatic control system, and the lack of sophisticated operator communication devices.

8.3.2 Computer Based Control System Design

8.3.2.1 Basic Concept

The experience gained from the design of conventional control systems can be helpful in designing more advanced control systems, but one must not be restricted by the limitations of conventional devices, otherwise, the experiences may become a liability, rather than an asset. In other words, one should never use an advanced control device, such as a programmable logic controller (PLC), just to duplicate the functions of relay logic devices. The product would inherit all of the potential problems of the conventional control system it mimicked. The important concept to consider when designing an advanced control system is that the system must be *trusted* by operators. Note that although all hardware components, including computers and PLCs, will ultimately fail sometimes, a computer based control system can be designed such that it will detect even its own errors, so its performance can be trusted.

For a new plant, an advanced control system design should be started from the field devices, such as valves and sensors, and all of the control devices should be integrated into one system.

8.3.2.2 Instrumentation Design

The guiding design concept for the field instrumentation system should be providing sufficient devices for error-free automatic control. Specifically, there should be redundant devices for every major measurement or actuating device, and feedback signals should be provided for every control action verification. This concept is very different from most conventional design concepts as can be seen from the following examples:

A simple instrument diagram of a remotely operated on/off type valve is shown in Fig. 8.7. As is shown, the valve can be either operated by a computer or manually, depending on the position of the computer/manual (C/M) switch.

Each valve is equipped with a fully open sense contact and a fully closed sense contact. When the valve is in any other position than fully opened or fully closed, then neither contact is closed.

When compared with the more conventional instrument design for remotely operated valves as shown in Fig. 8.8 (A) and (B), it can be seen that this design is considerably more complicated. The advantage of the new design is that the valve's position, whether it is open, closed, or in between, is clearly indicated. In addition, because of the multiple number of contacts, a single contact closure failure can be easily identified. For example, should both valve open and valve close contacts be closed at the same time, it is clear that one of the contacts must have failed.

A simplified diagram of a reactor solution charging system design is shown in Fig. 8.9. Note that two flow measurements and one level measurement are indicated. The solution charge tank has a uniform diameter permitting simple discharge flow rate calculations to be made based on the level change per unit of time. The total integrated flow

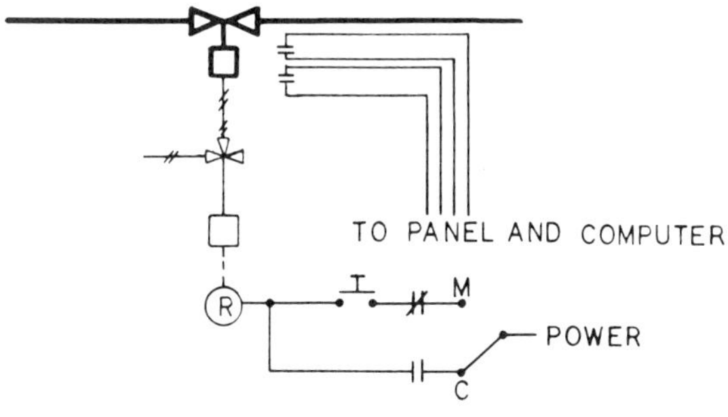

FIGURE 8.7 Typical control signal design for air operated solenoid valve showing computer, or a push button operation of relay operated air valve and open/close contacts for position verification signal to operator's panel and computer.

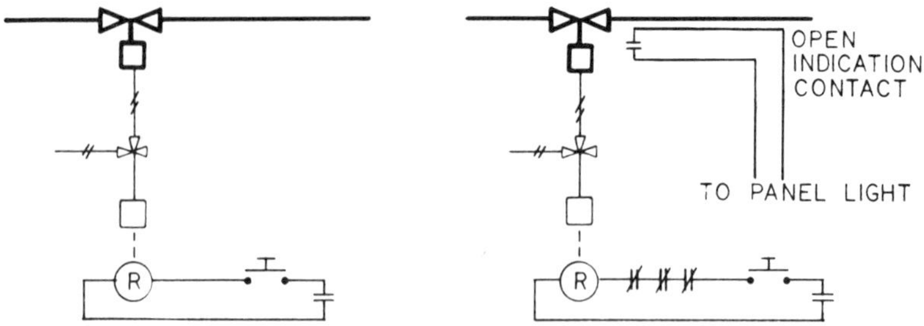

FIGURE 8.8 Typical control signal designs for push-button operated solenoid valves. The right design shows the relay-operated air valve being activated with a push button with three normally closed inhibit controls in series with the push button. The solenoid valve has an open indication contact connected to a panel light. If any of the three inhibit contacts is open, the valve cannot be activated.

can also be calculated based upon the total level change which was measured from the beginning to the completion of a charge operation. This design concept allows the computer to check the flow meter validity and perform corrections for solution temperature variations.

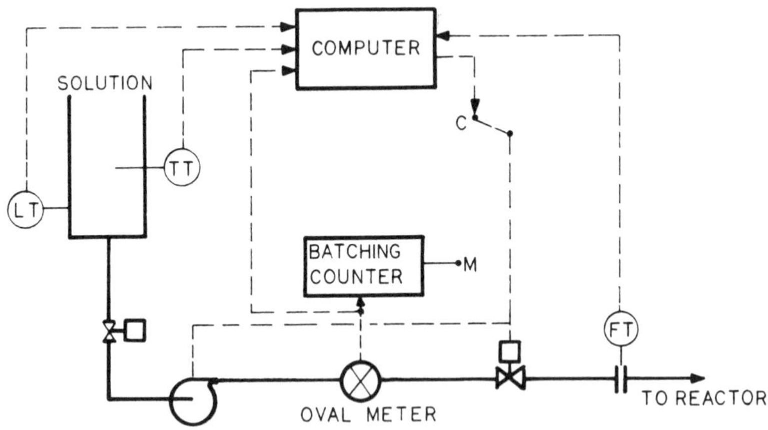

FIGURE 8.9 Computer-directed batch charging system with redundant measurements for charge quantity. If indicated charge rates as determined from level change, oval meter and orifice differ from each other by more than a predefined tolerance, charge is halted and alarm condition is indicated.

As shown in Fig. 8.10, all charge and discharge lines for each reactor are equipped with redundant valves. Each valve is independently controlled by the computer or by panel push buttons when the panel is in the manual mode. This design provides greatly increased safety and probability that the systems will remain on computer control even in the event of a valve failure.

In general this design concept, described above, is overly instrumented by conventional standards. However, with this degree of automation, all of the instruments serve specific purposes. Since computers are used to supervise these instruments, operators do not need to check them for validity or take any other actions. It is not overly instrumented by this standard.

8.3.3 Functional Specifications for Batch Control Systems

This section describes the basic functional level specifications for a batch control system, i.e., the major tasks to be performed by the control system. Users may use any type of hardware, such as computers, programmable logic controllers, digital controllers, or a combination of all types of hardware. There are indeed many ways to make a detailed design of the system, and the design possibilities will increase as new hardware and standard software systems become available. However, the functional level or preferred procedures for operating a plant do not have to change until entirely new process designs are developed, which does not happen nearly as frequently as electronics system designs change.

Fundamentally, a batch control system should be designed to satisfy the following conditions:

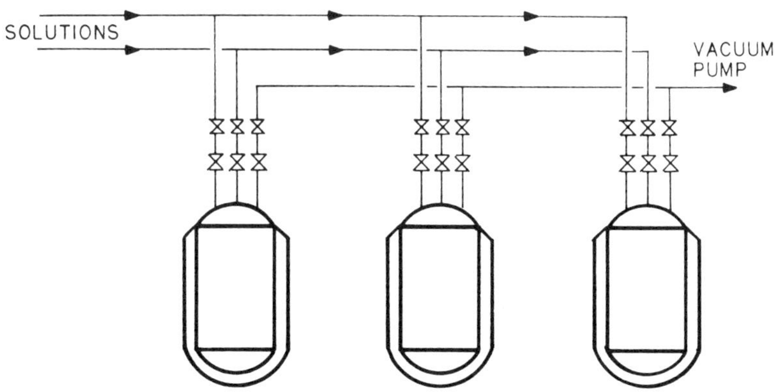

FIGURE 8.10 Double header valves provide necessary security to charge parallel reactors from common charging system.

8.3.3.1 Maintain Computer Control (Automatic) and Operator Control (Manual) Safety Separately

This extremely important precept for batch sequence control is sometimes ignored by many engineers, because it may not be important for a continuous regulation loop. An operator can switch a control loop to manual and change its output, and then switch it back to automatic without causing a problem for the control system because one control loop does not directly affect other control loop. An automatic batch control system combines the sequence control functions and the loop regulatory functions. It must "remember" past actions and frequently check the positions of all devices to assure that every device is at the correct position. A system that allows manual and automatic operation of the same device may have trouble in performing these safety checks.

One simple way to assure that automatic and manual control is safely separated is by specifying an auto/manual switch that controls all digital devices in an entire area, such as all of the header valves of a reactor. When this switch is turned to the auto position, all valves are under automatic control and visa versa. Figure 8.11 shows a permissive logic for selecting AUTO/MANUAL mode. Note that the "PERMIT FOR AUTO MODE" signal is set by a logic device, such as a computer, when all conditions for automatic control are met.

Operators may select the AUTO mode when this permit signal is on, usually at the beginning of a batch cycle. An operator may choose MANUAL mode at any time during a batch cycle by pushing the MANUAL button. He can then control all devices manually, but he cannot return to AUTO mode until the "PERMIT FOR AUTO MODE" signal is set again.

Maintaining automatic control and manual control separately does not mean they cannot be applied on the same process. For some

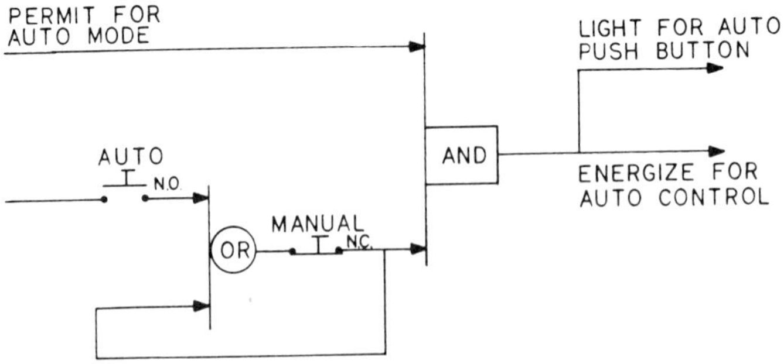

FIGURE 8.11 Interlock logic for permitting automatic control of batch sequences.

processes, it could be too expensive to install charge tanks and flow meters for every chemical to be charged into a reactor, or it could be impractical to test automatically whether a holding tank is ready to receive a discharge from a reactor. Thus, it is common to have a chemical charged to a reactor manually by using a charge bottle, or to check a holding tank connection manually. These types of operations can be integrated into an automatic control system. For example, a sequence control can be designed to reach a "hold" condition and inform the operator when a manually charged chemical needs to be added. The sequence control can be restarted either after the operator gives a proceed instruction, or by detecting a change of status.

8.3.3.2 Monitor Control Device Failures by Comparing Redundant Sensors

This is, again, an extremely important function. It is not a simple task because an equipment failure may cause many associated malfunctions which will result in many additional alarms. A reactor cooling-water pump failure during an exothermic polymerization reaction is a good example of this type of condition. This failure may cause a low flow alarm, a high temperature alarm, and a high pressure alarm. Since high temperature and high pressure alarms can also be caused by other failures, such as temperature controller failure, it is not an easy task for operators to identify which is the prime cause of the alarms. Note that a low flow alarm for cooling water may not be caused by pump failure alone; it can also be caused by control valve failure or by the flow sensor failure. For a critical variable, such as reactor cooling-water flow, it is very desirable to have redundant sensors, and generate alarm messages based on a logical evaluation of all related sensors.

8.3.3.3 Provide Clear Information Display for Every Interlock and Permit Manual Override

All interlock functions should have clear operator displays and permit manual override. Figure 8.12 shows a simple interlock logic for permitting a solution pump to start. All conditions connected to this "AND" gate must be met, otherwise, the pump cannot be started. When the pump fails to start, the operator will have to search a long list of instrument statuses to identify the reason for the fault, unless the control system gives him the reason by showing a clear message at the alarm typer or CRT. Conventional control systems, using relays for interlocking, usually do not give such a message. This is the reason that many of the interlocks in a plant are bypassed by operators. It is not uncommon to find jumper wires between terminals of interlocking systems. A clear message to indicate the reason for not

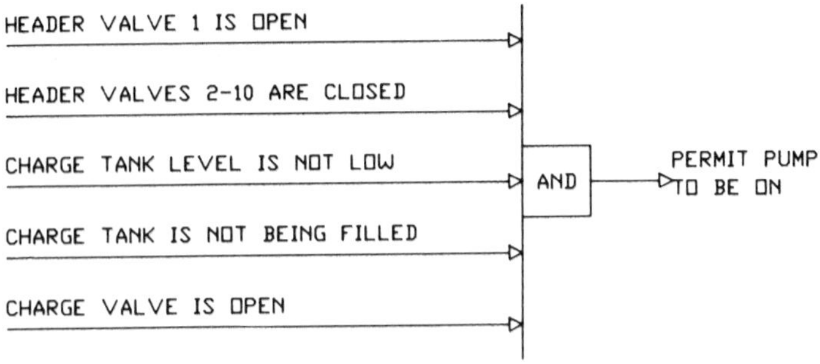

FIGURE 8.12 The permit signal for starting a pump depends on the conditions of many signals. When the permit signal is not on, a clear message is needed to indicate the reason for it.

permitting an operation and a manual override function will greatly improve the changes for interlocking logic devices to remain on line.

The reasons for allowing manual override of an interlock function need to be discussed. First, one cannot keep all interlock logic updated with all plant changes. When a device is temporarily disconnected, every interlocking function using that device will be affected. Rather than changing the design of all functions and then changing them back again later, it is much easier to override these functions.

Second, a "valve fails to close" signal may not mean just that. It could be that the valve has actually failed to close, but it could also indicate any of several other possible failures, such as a valve open contact failure, a loose wire, or a digital input card failure for the controller. When a "valve fails to close" is actually caused by a contact closure failure, it becomes more desirable to override this interlock logic and repair the contact closure later.

An override function must be carefully designed and managed. When an operator exercises an override, his action should be documented, and he must be fully aware that he is responsible for his action. Some operators may abuse the override capability and practically negate all safety interlock functions. Because of this, one may choose not to allow any override for super-critical functions. This type of decision should be made according to an individuals's desire and experience, and be clearly defined during the system specification period.

8.3.3.4 Test Feedback Status for Every Operation and Provide Error Recovery Sequence Control Functions

As described before, a valve should be equipped with a full open contact closure and full close contact closure, so an action by the control system will be confirmed by a change-of-status input signal. The logical steps for implementing such functions are quite straightforward when there is no error; normally these steps can be standardized and be applied to other similar devices. Figure 8.13 shows an "OPEN VALVE" subroutine for a computer-based sequence control system. When an error does occur, there could be many alternative responses to that condition. Figure 8.14 shows the simplest alternative, that is, wait for manual override. For a more complicated process, it is desirable to have a much more sophisticated automatic error recovery routine. These types of functions have to be designed based on an in-depth understanding of the process and the operating procedures.

To illustrate the complexity of an error recovery function, consider the process shown in Fig. 8.1, and the time chart in Fig. 8.2. Now, assume both Chemicals 1 and 2 charges have started successfully and are being charged, and it is time to start charging Chemical 3. The control system then detects that the charge pump for Chemical 3 cannot be started, so the sequence control is branched to an error recovery routine. This is shown in Fig. 8.15. Note that when the chemical charge pump fails to start, it is necessary to stop the charging of both Chemicals 1 and 2. Otherwise, it would

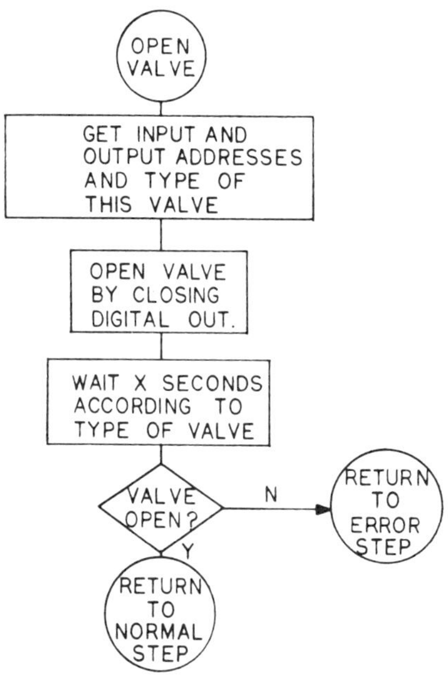

FIGURE 8.13 A common open-valve subroutine with error detection for different types of valves.

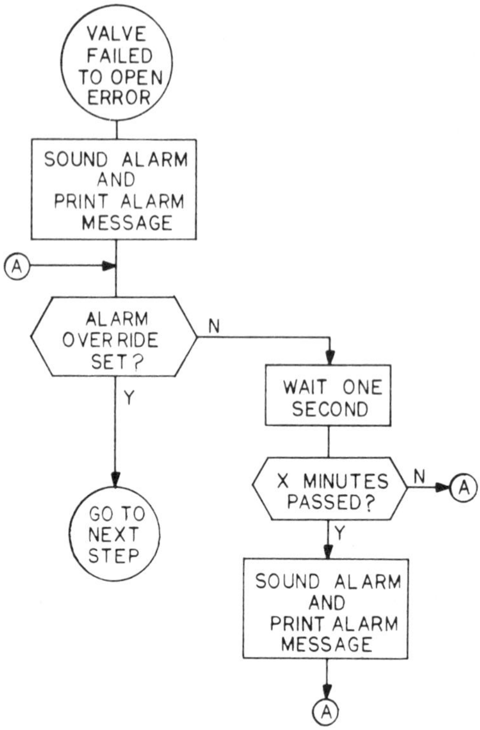

FIGURE 8.14 The simplest error recovery routine waiting for manual override.

be difficult to charge Chemical 3 at a later time because the charging of Chemicals 1 and 2 would be completed before the operator can repair the Chemical 3 charge pump, and Chemical 3 should be carried into the reactor by Chemical 1. As shown in Fig. 8.15, this error recovery routine will provide a method to restart the entire charging operation automatically after the charge pump for Chemical 3 is repaired.

It is clear that when an engineer wishes to automate error recovery routines, he must be prepared to develop very sophisticated and lengthy sequence control functions. There is no way to give a generalized figure of magnitude or how involved an automatic error recovery routine might be, but it is safe to say that it can be more than 10 times more complicated than the routine sequence when everything is normal.

There are easier ways to specify an error recovery routine. One is to switch everything to manual and complete the charging by manual control. This will require well-trained operators and adequate information displays when the process is stopped as a result of an error. Another method is to dump Solutions 1 and 2 already charged to the reactor manually and restart the automatic sequence at the beginning of the cycle. This approach is simple, but will be costly in lost time and also in the cost to separate Chemicals 1 and 2.

8.3.3.5 Minimize False and Unnecessary Alarms

It is very easy to experience false alarms for a batch process because of the transient nature of the process. For a continuous process, one can implement a deviation alarm, that is, when the deviation between the setpoint and the process variable exceeds a limit, for a control loop, without worrying too much about false alarms,

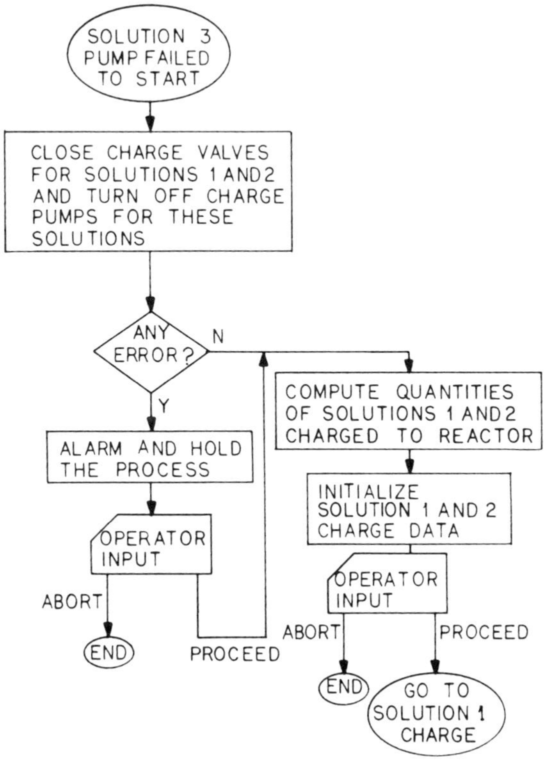

FIGURE 8.15 Error recovery routine for solution charging sequence; a special routine is needed for very process.

because the control loop is supposed to operate within a given range. A deviation alarm for some batch processes can have false alarms every cycle. For example, a reactor temperature may have a profile as shown in Fig. 8.16. When a normal deviation alarm function is implemented for that temperature loop, there will be an alarm during heat-up and an alarm during cool-down. This type of false alarm disturbs operators and will make them less responsive to alarm conditions. To correct this problem, one can use the sequence control functions to suppress the alarm before the heat-up and activate the alarm when the setpoint is reached, then to suppress the alarm again at the end of the cycle.

It is also important to remember that an alarm should only be generated when it is necessary to bring some information to the operator's attention, most likely, when the operator's actions are needed. It is a common mistake made by many engineers to generate many unnecessary alarms or computer printouts. Past experience indicates that false alarms and unnecessary alarms may cause operators to ignore real alarms too; thus, meaningless alarms must be minimized.

8.3.3.6 Automatic Processing of Recipe Information

As discussed before, recipe information for a batch control system can be stored in a computer. This information should be automatically transferred to the sequence control function at the time of need. For example, the charging quantity of a chemical should be set automatically according to the selected recipe number. Conceptually, all an operator needs to do about a recipe is to check whether

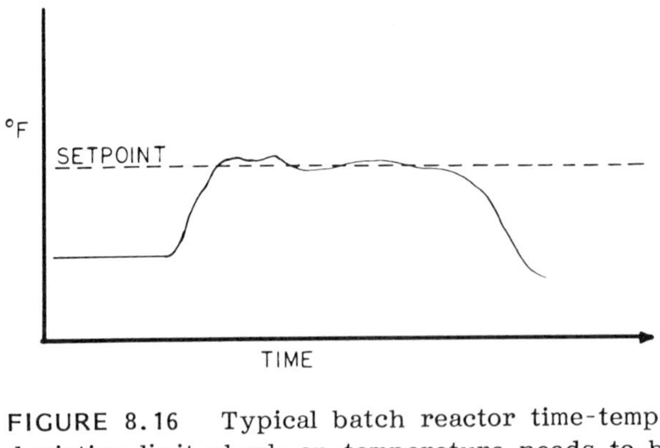

FIGURE 8.16 Typical batch reactor time-temperature profile. A deviation limit check on temperature needs to be suppressed during both heating and cooling periods, or else false alarms would be generated.

all the numbers are correct in the computer and to enter the recipe number in the sequence control function. He should not need to enter any values during a sequence.

There are many good reasons for the implementation of automatic recipe processing. The most obvious benefit is saving time. When recipe information has to be manually set, many automatic sequence control functions cannot be optimized. For example, a delay timer for a sequence control may be set according to the recipe information (some products may require longer a time delay than other products), thus, minimum delay time can be expected for every cycle. When the recipe information could not be transferred to the sequence control, the longest delay times would have to be used for all recipes, thus, prolonging the cycle time for some products. Of course an operator can adjust timers manually, but when an operator has to enter a large number of variables into a sequence control system, he will have to read a long list of variables and make many manual adjustments. This will prolong cycle time also. The most important benefit for automatic recipe processing, however, may be the elimination of data entry errors, which could easily happen when a large number of recipes are used. The potential hazards for entering wrong data could lie anywhere between bad product quality and a major accident.

8.3.3.7 Monitoring and Responding to Process Emergencies

The monitoring of emergency conditions for a batch process, like all other functions, is also more complicated than that for a continuous process. Figure 8.17 shows a reactor with a water jacket. When this reactor is used in a continuous polymerization process, i.e., material flows into and out of the reactor continuously, a failure of the circulating pump will always be an emergency condition, because the loss of cooling water circulation will heat up the reactor which may become explosive. Thus, a low pressure or low flow rate for the cooling water can be used to trigger an emergency alarm and actions, such as to stop the chemical reactions. For a batch process, a reactor may be used for different purposes depending on the time of the cycle. A batch polymerization process can use a reactor for polymerization first, and then as a stripper to remove the unreacted monomers. A loss of the water circulating pump during polymerization is an emergency condition, but a loss of the same pump during the stripping operation is not an emergency condition. The control system should be able to tell the differences between these two states to assure that emergency conditions are identified without false alarms.

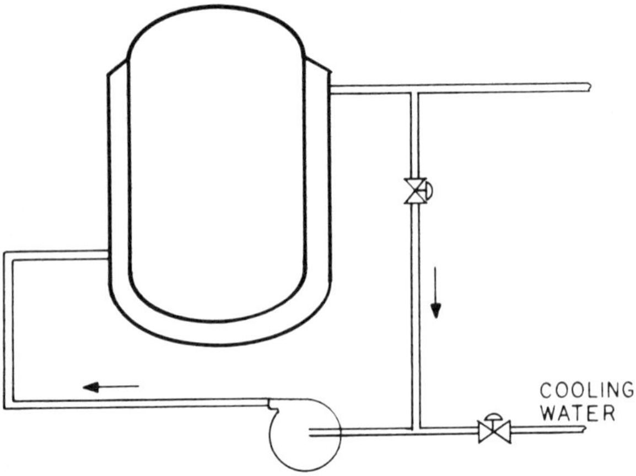

FIGURE 8.17 The failure of the circulation pump is an emergency condition only during the polymerization phase of a batch cycle.

8.3.3.8 Provide Advanced Control Algorithms for Reactor Control

In order to achieve the desired performance for reactor temperature control as specified in previous sections, a control algorithm more advanced than the conventional proportional-integral-derivative (PID) controller may be needed. One such algorithm is the extended PID algorithm as described in Eq. 7.64, 7.65, and 7.66 in Chapter 7. To implement this control algorithm to control the temperature of a particular reactor, the switching parameters and tuning constants in that algorithm must be set according to a predetermined sequence. This means sequence control functions must communicate with temperature control functions to a very detailed level.

8.3.3.9 Save Operating Data and Generate Performance Reports

One of the most common mistakes made by people inexperienced with computer applications is the tendency to save too much data and produce too many reports. When this is the case, the large volume of reports becomes useless because the task of extracting any meaningful information from them becomes too difficult. A good batch report is required for the performance evaluation of a batch process. The basic criteria is to save the significant information that affects the reactor throughput or product quality. The following list gives some examples:

All charge quantities as solutions and pure compounds
Batch start time
Batch ended time
Reactor temperature profile sampled every few minutes
Calculated conversion
Total heat of reaction
Time of process stoppage and reason
Alarm summary

8.4 CONTROL SYSTEM CONFIGURATIONS

Whenever batch control systems are discussed, many people tend to conclude that the control system will be based on programmable logic controllers (PLCs). Since many PLCs are now designed with sequence control capabilities plus PID algorithm and continuous current outputs, it is quite understandable how such a conclusion could be reached. In fact, many simple batch control functions can be easily implemented by using PLCs alone. Automatic soot blowing for a boiler based on a timer is such an example. This type of process does not have recipes, and program changes are rarely required. There is little need for a sophisticated operator communication function, or any computation function. A PLC can be configured to control this type of process and give excellent results.

For controlling a more complicated batch chemical plant, however, a PLC alone is not sufficient. The major reasons are that PLCs lack the capabilities for handling the important functions such as automatic recipe processing, sophisticated operator communications, or advanced control algorithms. For implementing the control system specified in the previous section, there can be three hardware configurations based upon different design criteria as described next.

8.4.1 Distributed Control System with Batch Controllers and Host Computers

This is probably the most desirable control system configuration for a new plant because this type of control system can take advantage of all the latest technology, such as high reliability and low installation costs as described in a previous chapter. Figure 8.18 shows a simple block diagram for a batch control system.

There are several important hardware specifications one should consider before selecting such a system. Note that not every distributed control system supplier can meet these specifications, therefore, it is very important to choose a system that has sufficient capability for the job.

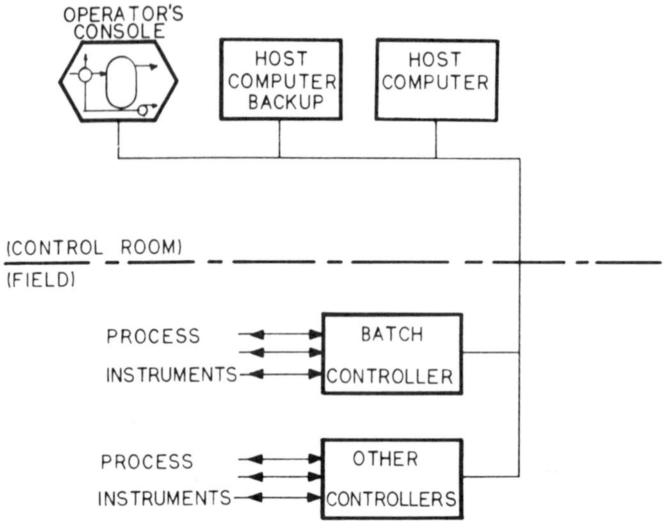

FIGURE 8.18 Control system configuration for a batch chemical process using redundant host computers supervising advanced digital controllers.

Redundant host computers may be needed when information handling is very important, such as for a plant producing many product grades. When redundant computers are used, make sure that the backup computer has the means for obtaining the latest process information in case the on-line computer fails.

Communication between the batch controller and other controllers is needed. Suppose a batch controller is used to handle sequence control functions, plus simple loop control functions, and another digital controller is used to implement an advanced control algorithm, then these controllers should communicate to each other to transfer control information, such as change of control mode, etc.

Operator's console should have sufficient capacity to handle manual control for all devices, and to satisfy the criteria of maintaining automatic control and manual control safely separated.

8.4.2 Redundant Computers With or Without Manual Backup

Many batch control systems have been implemented using redundant computers directly connected to the process without controllers.

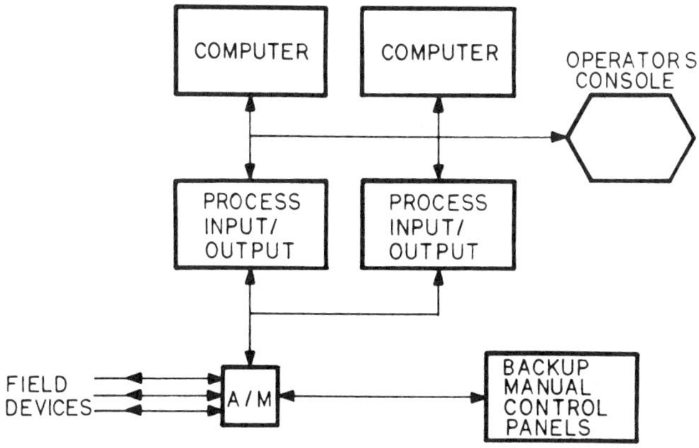

FIGURE 8.19 Control system configuration for a batch chemical process using redundant computers directly control the process.

The major difference between this concept and the previous one is the design of a manual backup system. Note that manual control using a distributed control system is not the same manual mode by conventional definition, which implies operating field devices by push buttons, or potentiometers directly. When manual control is performed using a distributed control system, operators still depend on the computers that run the CRTs, push buttons, and communication devices.

Some people may not feel comfortable about this design concept and prefer to use real manual operation. Figure 8.19 shows a block diagram that describes a control system using two computers, two input/output subsystems, and manual backup panels. Note that the auto/manual switch only controls the control system outputs. When it is turned to auto, the computers control the field devices; and when it is turned manual, the panel controls the field devices. All measurements are connected to both systems all of the time.

The design, fabrication, installation, and testing of backup manual control panels is a costly and time-consuming task. It even increases the complexity of the software design of the automatic control system, but manual backup has been used in many installations.

Regardless of how reliable a redundant computer system might be, it is not 100 percent reliable and, therefore, some form of backup is necessary for most chemical processes because of safety considerations. When complete manual backup is too expensive, it is possible to design a limited backup system just sufficient to shut down the plant safety when the computers fail.

One of the most important advantages for the use of a computer only approach is that all functions are handled by the computer,

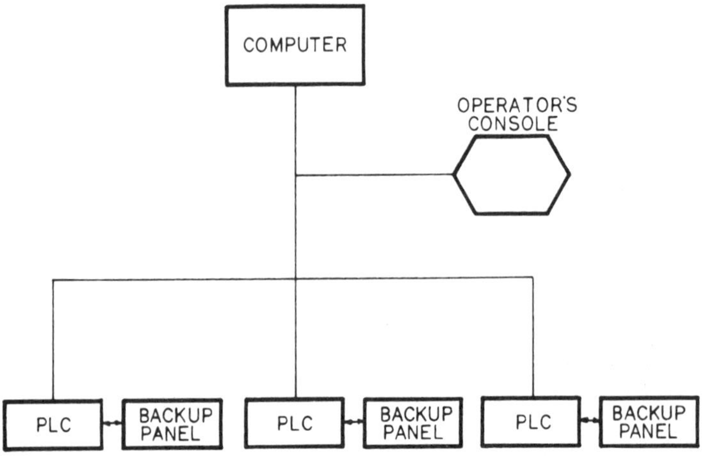

FIGURE 8.20 Control system configuration using a single host computer supervising PLCs.

thus, communications between functions are not a problem, provided adequate software is available. Then, the most important consideration for using a computer only approach is to choose a system with proven batch control software. Many computers vendors claim that they have such a software system, but users must evaluate whether there is sufficient software to implement all desired control function, and to interface with backup hardware systems. No one can have a complete system which is "the best" for everyone. A well-conducted engineering evaluation of all potential systems before a project is started will save a lot of trouble later.

8.4.3 Computer with Programmable Logic Controllers

Figure 8.20 shows a control configuration utilizing a computer plus a number of programmable logic controllers. Note that during the normal operating mode, the operator can control the process from the computer-driven console. When the computer is down, the operator can control the process from the backup panels driven by PLCs. If a PLC should fail, it affects only part of the plant, the remainder of the plant would not be affected, so it may be an acceptable risk. This configuration may not require complete manual backup for this reason.

As discussed earlier, PLCs do not have sufficient power to implement advanced control algorithms. In addition, some PLCs on the market do not have redundant data links to computers, and the communication devices at the PLC level can be very limited. The configuration shown in Fig. 8.20 may be only suitable for simple processes.

REFERENCES

1. Nakagawa, Mitsuru, Large PVC manufacturing plant and computer control system, *Chemical Economy and Engineering Review* Japan, March and April, 1970, pp. 12–16 and pp. 27–32.

2. Billmeyer, Jr., Fred W., *Textbook of Polymer Science*, Interscience Publishers, New York, 1962.

3. Flory, Paul J., *Principles of Polymer Chemistry*, Cornell University Press, Ithaca, New York, 1953.

4. Albright, Lyle F., *Processes for Major Additon-Type Plastics and Their Monomers*, McGraw-Hill, New York, 1974.

5. Luyben, W. L., Batch Reactor Control, *Instrumentation Technology, 22*(8):21–24 (August, 1975).

6. Amrehn, H., Economic incentives for computer control of polyvinyl chloride plants, Instrumentation Forum for the Petroleum and Chemical Industries, Foxboro, Massachusetts, June 6–11, 1971, pp. 1–13.

9

Supervisory Control Systems

9.1 THE EVOLUTION OF SUPERVISORY CONTROL SYSTEMS

When digital computers were first used for on-line control applications in the 1960s, most of them were installed to enhance existing centralized control systems with a large number of analog controllers. These enhancements added more "intelligence" to the control systems, so the operators could use these systems to maintain the process at a higher performance level than that achievable by using only analog controllers. These types of computer applications are called supervisory computer systems (1–3).

The hardware configuration of a supervisory control computer has been discussed elsewhere (see Chapter 5) and will not be repeated. These earlier systems were expensive because of high hardware costs and the need for extensive software development for every installation. A detailed feasibility study was required to identify the cost and benefits before a project could be supported by management. However, past experience shows that after a project was completed, many of the benefits identified during the feasibility study would not be realized, while many side benefits or unexpected results helped to eventually justify the project economically (4). Some projects were reported as "technical successes, but economical failures," which implies that the cost of the project could not be recovered.

The most important contribution to the future of process control derived from these early supervisory control systems, whether they were economical successes or failures, was the technical know-how which they provided on the application of computer systems for on-line process control.

The wide acceptance of distributed control systems since the late 1970s has made the implementation of supervisory control much easier and less costly than ever before. As shown in previous chapters, a supervisory control computer can be connected to the communications network of a distributed control system and communicate directly with the data base. This eliminates the need for interface hardware and, thus, greatly reduces the supervisory control computer hardware cost.

The availability of high level programming languages and advanced operating systems has made the software development task easier than before. In addition, generalized supervisory control software packages are now offered by many vendors and software system houses. (These will be discussed later in this chapter.) Software development tasks are now manageable in terms of cost and schedule (5).

With lower costs for both hardware and software, will the supervisory control system be considered as a low budget item which requires no detail cost/benefit analysis from management's point of view? The answer is maybe, but unlikely nor desirable. The fact is, today's large supervisory computer installations are many orders of magnitude faster and larger in memory capacity than early generation computers, and they are designed to provide much greater functionality than ever before, so the supervisory control computer system can cost even more than a computer purchased 20 years ago. On the other side of the spectrum, one can purchase a small computer and use it within a limited scope to control a process plant. This will be a low-cost computer, but the benefits will also be limited. In conclusion, the cost of supervisory computers today can range from a few thousand dollars to a few million dollars. A feasibility study is still needed for most applications, because even a low cost computer may require extensive software development, so the system with the lowest hardware cost may not be the best system.

The latest application concept in supervisory control is to combine several major computer functions, such as process control, on-line performance/cost analyses, and management information decision support into a network which consists of many computers (6). Figure 9.1 shows a configuration for such a network. Note that several supervisory control computers and a "plant management system" are connected together to form an information network. The major goal for such a system is to optimize the plant's performance from both the technical and managerial points of view. The availability of realtime operating information to management may have a major impact on plant operating philosophy and even future plant organization. Some middle managers may no longer be needed to interpret

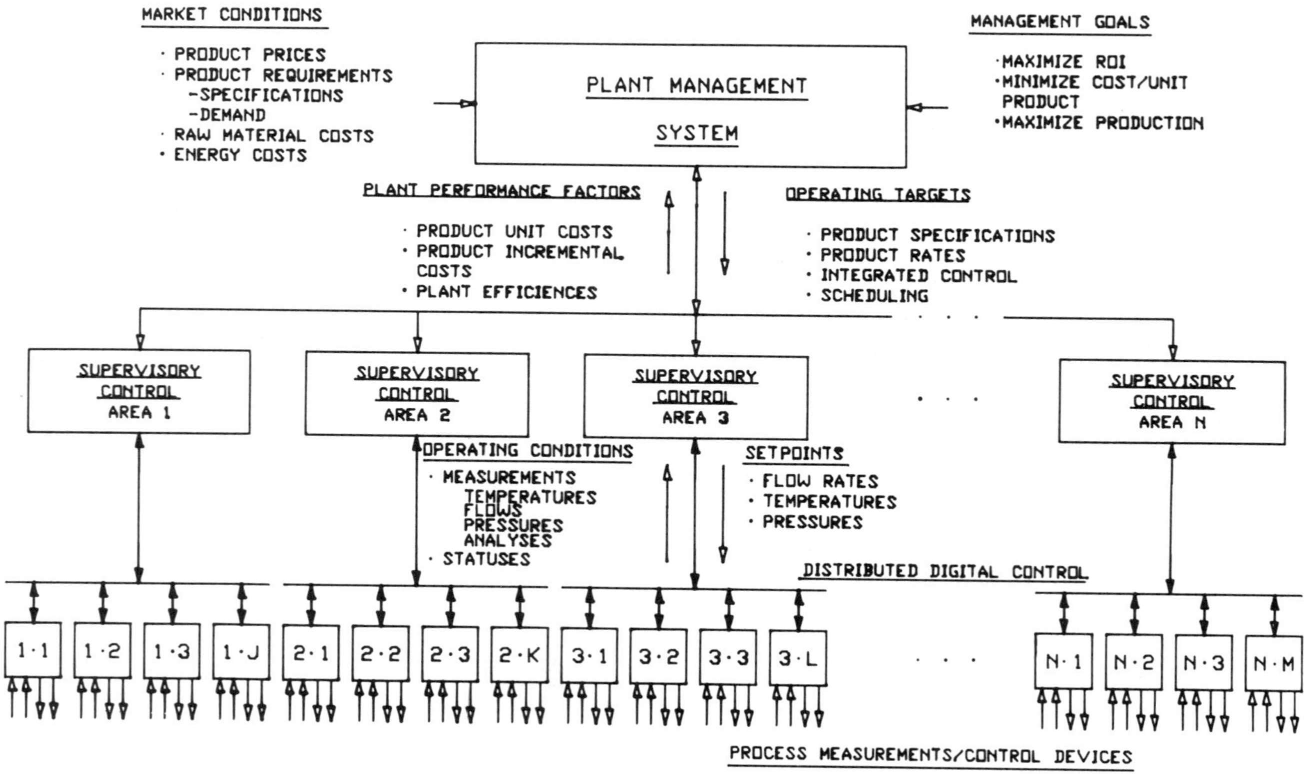

FIGURE 9.1 Block diagram of a multitiered plant integrated control and management information system showing classifications of information flowing between tiers.

plant operating data for senior managers, nor will they be needed for transmitting senior management decisions to plant personnel; all of these functions can be accomplished automatically. History has proven that fewer managerial levels improve overall effectiveness, so the potential savings from this type of system can be enormous. A manager now faces two questions; can he afford to have a computer, and, can he afford not to have a computer?

9.2 IMPORTANT FACTORS FOR A SUCCESSFUL SUPERVISORY COMPUTER SYSTEM

The term "supervisory computer" is used to represent all computer systems residing above basic regulatory loops whether they are implemented in either the analog or digital form. In actual installations, there could be more hierarchical levels than that shown in Fig. 9.1. There could be a network of special purpose supervisory computers for certain areas of a plant which communicate to another supervisory computer for overall coordination.

Figure 9.2 shows an example of a computer system network for a pulp and paper mill. The mill management system is used to collect process data and convert them into useful information. This information can be used to support decisions such as the scheduling and load allocation of each machine to achieve the best economical results. For each paper machine, there can be special function control computers such as basis weight, moisture and speed control systems, and there are control functions performed by a distributed control system and programmable controllers. The main reason for this rather complicated configuration is that many special purpose computers have been implemented over the past years, and they cannot be replaced by a single vendor source. With such a broad scope, how should an engineer, or a manager start a project? The following paragraphs describe a general guideline.

9.2.1 People and Organization

The most important factor for the success or failure of a computer project, or for any other business venture, is to find the right people for the job. It is particularly important for a computer project because many managers do not recognize the right type of people needed. A common mistake in the past has been to use a computer programmer who has no knowledge about the process, or to use a project engineer who has no knowledge about computers, nor much about the process, as the project leader (7). This mistake has certainly given a bad name to many projects over the years. A project leader should be a person who has solid experience in computer

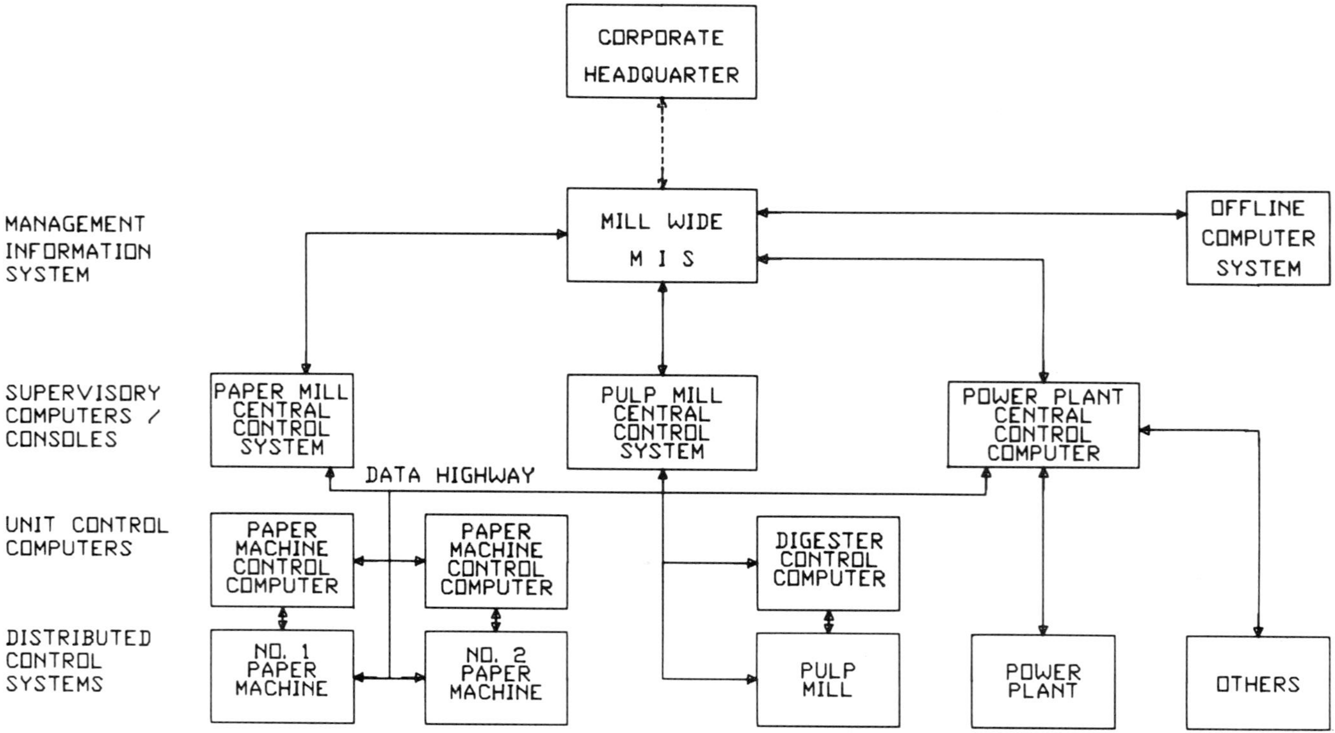

FIGURE 9.2 Integrated computer control system for a pulp and paper mill including distributed process control systems interfacing to local supervisory control systems which are, in turn, interfaced to the mill-wide management information system.

systems applications and in project management. This person should have an engineering background and must understand *both* the process *and* the control system including the computers, and should also be able to maintain a clear view of the project objectives. It is important for industrial managers to either develop such type of people within their company or to look for experienced people from outside (8).

The type of technical expertise required to implement and/or maintain a supervisory control project depends very much on the type of project. A turnkey project may not require any control system expertise from the user company, while an in-house developed integrated system may require several people in each discipline, such as instrumentation design, process analysis, control system analysis, computer hardware and software design. Generally speaking, manpower requirements also change with technology. A large project team formed before the mid-1970s required a large number of computer programmers because many special programs were developed by the user, and some programs had to be written in low-level (machine) languages. The availability of high-power computers at low-cost has made "user friendly" software packages possible, thus, the number of professional programmers needed to implement a project has been greatly reduced. However, the more complicated hardware configurations of today's systems requires more system design and digital communication expertise than needed for implementing a similar-size project of some years ago. With the continuing fast-paced development of electronic technology, the need for special technical expertise required for future projects will change, and users will need to develop this expertise in order to take full advantage of new technology.

Just how many people are needed to implement a supervisory control system at this time? Based on the authors' observation, the number can be anywhere between two to over 20 people, depending on the complexity of the process. The project team for an ethylene plant supervisory control system completed in 1982 was reported as 19 people from user, computer supplier, and consulting firms (9). The seemingly high number of people involved in this project is an indication that many specialists were needed at that time. Should a similar project be started in 1985, a somewhat lesser number of people would be needed because of the advances in computer systems as discussed before, but the tasks such as the analysis of process performances, the design of sampling systems for analyzers, or the training of operators, etc., will still be manpower intensive. In fact, it is not uncommon to find that the manpower cost for implementing a supervisory control project exceeds that for purchasing the entire hardware system. It is important that management recognize this need before starting a project and keep this team together during the entire duration of the project.

One important consideration concerning process control project personnel is where to put them. Many computer related people are located in research and development departments or in "EDP" departments. This is a mistake. Supervisory computers are productivity tools, and people who are responsible for these projects should be part of production departments. They should be a staff group to support the production department, and their responsibility should be clearly defined. Staff people are responsible for technical support and coordination between plants, while production people are responsible for the overall performance of the computer system (10).

9.2.2 Establish a Long-Range Project Plan

When the final design goal is a plant-wide control system consisting of a number of computers in hierarchical form, three important considerations arise: (1) the priority of installation, (2) the communication, or interface, between the systems, and (3) the maintenance for both system hardware and software. Before any important decision is made, such as the selection of a computer vendor, all three factors should be carefully evaluated. The low bidder for a project may not be the best vendor when the whole picture is considered.

9.2.2.1 The Priority of Installation

Based on the experiences of existing installations, the best way to start a project is to build a system from the bottom level and grow to the next level one step at a time. This means that one should start a project by implementing regulatory, or basic, control functions, followed by first level supervisory control functions, followed then by optimization functions and management information systems. This approach has many advantages. The most important advantage is that by implementing the regulatory functions first, the process will be stabilized, and the cause-effect relationships between process variables can be more clearly identified. In addition, regulatory control functions are simpler, and their benefits are more obvious, thus, they can be easily accepted by both management and operations people.

Special purpose computers, such as a quality control laboratory computer, and chromatograph computers, should be installed prior to the completion of the main supervisory system. During the early years of supervisory control computer applications, many users installed computers to perform on-line optimization for a complicated plant. The implementation procedures at that time were to first develop sophisticated mathematical models for the process and for the optimization calculation and then develop special computers programs to implement these models. While there were successful projects at that time, the requirements for technical expertise and project costs were high, and, thus, became prohibitive for many companies. In addition, one of the most difficult tasks for process modeling development is to collect accurate data. A well-designed computer based system for regulatory control will greatly reduce the complexity of this task and improve the overall effectiveness for developing process models, so the optimization functions can be implemented with less manpower cost.

9.2.2.2 The Interface Between Computers

Whenever the interface between computers is discussed, many people suggest standardization as the answer. In reality, on-line computer products are not standardized and probably will never be. One can argue that the RS-232 digital communication is an industrial standard, but even two RS-232 compatible devices may require special software work to communicate. In addition, when a large quantity of information needs to be transmitted, the rate of RS-232 may not be fast enough, and the high speed data communication protocol of one vendor is rarely compatible with that of others, at least not at this time.

Without standardization, the next easiest answer is to assure compatibility by buying systems from the same vendor. This may or may not be a good solution. The major problem is that even the biggest vendor may not offer everything required to fit the user's needs. In addition, rapid technological development always encourages new vendors and makes old ones obsolete. To stay with one vendor could be a costly mistake when that vendor is not a progressive company.

The latest communication devices offered by control systems vendors are the so-called "universal gateways" which are supposedly able to interface any desired hardware to a vendor's communications links. There is not yet sufficient information about these devices, but the cost of these devices can be high, and special software development may still be necessary.

One good way to assure system compatibility is to make a detailed engineering evaluation case by case. The best method today may become obsolete by next year. This reconfirms the discussions in the last section about the importance of having the right people at the job. The final goal is that all computers must be compatible, but one must study the individual requirements to decide how to accomplish this goal.

9.2.2.3 Maintenance

The scope for maintaining a computer system is different than that for maintaining most other major equipment. To maintain sophisticated equipment such as a compressor, the major concerns are spare parts inventory and the training of maintenance technicians. In other words, "hardware" maintenance is a major concern. For a large sophisticated computer system, a hardware maintenace contract can be purchased from the vendor. For a smaller system, self-diagnostics can pinpoint most hardware problems and any maintenance technician can perform a card replacement task. The maintenance of computer hardware, thus, becomes straightforward from the user's point of view. However, the maintenance of software is not as straightforward. For most computers, there is a continuous evolution of basic software (this is usually done by vendors), but these changes must be made to all existing systems. In addition, application software, whether developed by user or vendor, must be updated when plant changes are made. Last, but not least, all changes must be well documented. The maintenance of software requires disciplined technical people, and these people do not fit the traditional type of maintenance department. How to organize such a group and to use them effectively is a challenge and rather subjective. Like most managerial questions, there is no scientific measurement of a right or wrong answer, only results over a long time will tell. It is the authors' opinion that software maintenance should be performed by the same group that implemented the system, and the cost for supporting such an organization permanently should not be overlooked when a project is planned.

9.2.3 Define Functional Specifications for the System

The task of writing down a functional specification is extremely essential to the success of any project. The specification provides a common understanding of the scope and expectations of the project, which can be reviewed, debated, revised, and ultimately approved by all parties concerned. If the specification is well thought out and documented, then hopefully, the project is one giant step closer to a successful completion.

Without an adequate functional specification, there may be those on the fringe of the project or those who will benefit from the project upon its completion, that have their own understanding of the purpose of the project. This may be at odds with the understanding of the project implementor and, therefore, the risk of conflicts and additional costs associated with scope changes may sour management's perception of the entire project.

Specifically a functional specification is needed to describe: (1) what are the tasks the computer is expected to accomplish, (2) how is the computer to accomplish these tasks, and (3) what resources are needed to accomplish the desired tasks? Specifically, the following subjects need to be clearly specified.

9.2.3.1 Define Advanced Regulatory Control Functions

As discussed before, advanced control functions should be implemented using a bottom-up approach, that is, implement the simple regulatory control functions first, and then evaluate the potential benefits for optimization functions after the regulatory functions are completed. Advanced regulatory functions require neither sophisticated process or economical models, nor optimization techniques, but the benefits are well known. The type of advanced regulatory control functions and the techniques needed to implement them have been discussed before and will not be repeated here. The following information about each advanced control function must be specified:

Input process variables, including on-line continuous measurements, analyzers, and laboratory inputs.
Computation equations for computed process variables.
Control equations.
Initialization procedure and failure mode analysis.
Control output locations.
Special operator communication requirements, such as special alarm messages.

One of the most important guidelines for developing functional specifications is to design the control functions for easy use from the operator's point of view, that is, to make the control system as user friendly as possible. Note that this requires more than just better operator communications functions, which is discussed later. The basic requirement for a user friendly supervisory control function is to design it in such a way as to permit single-button implementation by the operator with the computer testing all permissive conditions automatically. When a supervisory function cannot be started for of any reason, the computer should give a clear message stating that reason. An example for an automatic initialization function is examined in Appendix A. The manpower cost for implementing this type of function is high and must be allocated before a project is started. However, it can be easily justified because when a control function requires excessive manual data entry by the operator, that function will not be used much.

9.2.3.2 Define Monitoring Functions

Most process monitoring devices can only read and test one single process variable, such as sounding an alarm when a temperature exceeds a predetermined value. A computer can be programmed to perform more sophisticated monitoring functions, such as to sound an alarm when a column is close to flooding or the flare loss is becoming too high.

One important guideline for specifying the process monitoring function is to minimize the chances for "false alarm" conditions. This may not be as easy a task as some people expect because in order to minimize the chance for a false alarm, the status of all measurements needs to be considered in the monitoring logic. This makes the implementation more difficult, but also improves the operator's confidence of the monitoring functions. Several system suppliers have developed so-called expert systems or artificial intelligence to help operators monitor processes (11,12). More such systems will likely be developed in the future.

9.2.3.3 Define Performance Calculations

The difference between the task of defining performance calculation and the previous task is that the results of performance calculations do not require immediate operator attention, consequently, the operator communication requirements are different. The results of the performance calculations are shown in displays or reports, but do not sound an alarm. A simple example is the computation of heat transfer coefficients for heat exchangers. The results from this calculation are used for plant efficiency improvements, so they must be as reliable as those for the monitoring functions.

9.2.3.4 Define Operator Communication Functions

The operator communication functions for a supervisory computer are more complicated than many people anticipate because they must be integrated with the control and monitoring functions as discussed in previous sections. Because of this requirement, these functions must be specified clearly during the early stage of a project in order to estimate project cost.

The basic concept for specifying the operators' communications functions is to provide plant operators with sufficient on-time and clear information during normal, abnormal, and emergency conditions (13). In addition, operators must be able to instruct the computer with fast and simple actions. This can be a time-consuming and costly requirement, but can also be necessary. For example, the most commonly used operator communication device is a CRT with a keyboard. Since a modern supervisory computer has a large memory capacity, it is important that these displays are organized according to a well-developed pattern so that operators can easily locate the displays as needed. One way to accomplish this is to design an overall graphic display for a process unit with the capability to zoom into a detailed display of all control functions. When an emergency alarm occurs, the information related to that alarm should be displayed by a single request at the keyboard. It is possible that more than one CRT display is needed during an emergency condition; however, one must be careful not to give too much information to an operator and confuse him. This makes the specification of operator communication functions challenging.

Defining all displays will take time and may not be practical at the time of system specification, but a reasonable estimate must be made as to how many displays will be needed and how complicated these displays will be. This information is required before the selection of a computer because the larger the number of desired displays, the larger the requirement for computer memory, and the higher the cost for the project.

Other operator communication functions, such as printing alarm messages, reports, etc., should also be specified by people with plant operation and control system experience. One should always keep in mind that the printer in a control room is very different from the one in a computer room. The number of printed messages should be limited to those concerning operators. Generally speaking, the number of printers should be minimal.

9.2.3.5 Define the Historical Data Storage and Retrieval Requirements

Historical data can be easily stored in computer data files, and these can be retrieved at any time for display and analysis. In fact, many supervisory computer system vendors offer "standard" historical data storage as part of their standard software package. These packages may meet the needs of many users, but some special processes may be required for data requiring very frequent sampling over a very long period of time, which will require special techniques to compact the data in order to be practical. When this is the case, a special software package will be needed and must be specified before the overall software system is configured.

As for data retrieval, there are many different methods, but the main difference is whether these data need to be used only for the on-line computer or will these also need to be used by other off-line computers. The requirements in hardware configuration and in software packages are quite different, thus, the need must be determined before a system is purchased.

9.2.3.6 Define Process Models and Optimization Techniques

Optimization has different meanings to different people. A single-variable control loop can be optimized by selecting the best tuning parameters based on a given performance criteria. For supervisory control, optimization usually means the identification of the operating conditions for a plant (or part of a plant) at which the highest profit can be made at any point in time. Note that an on-line optimization is needed only when the optimum conditions change frequently, otherwise, an off-line computer would be sufficient to solve the problem. As discussed elsewhere in this book, many processes may not need an on-line optimization function because they are operated at equipment or quality constraints. For some processes, the benefits of the optimization function can have a major impact on the plant's profitability (14). So the candidate for optimization must be carefully selected because the costs in technical expertise and computer systems could be high, and they should be justified.

To define an optimization function, one must have the following mathematical models:

The Process Model: For most applications to date, a steady state model is sufficient. This model is usually much simpler than a process design model, because the equipment design is fixed and the major concern is the cause-effect relationship between the independent variables that can be manipulated by the control system and the dependent variables that affect product quality or operating cost.

The Economic Model: This is a statement of the performance criteria or the object function which needs to be maximized or minimized. The most commonly used object function is the profit model of the plant, that is, the total value of the product minus the total cost.

The Process Constraints: The equipment and product quantity and quality limits which affect the operation.

The Optimization Model: The mathematical package that is used to find the optimum conditions, such as linear programming, gradient search, or Lagrangian multiplier method. These techniques were developed long before the large-scale operation of computer systems as shown in many text books (15,16). Unlike computer systems, optimization techniques for supervisory control have not changed that much from those text books, as can be seen from recent publications (17).

Some popular process models, such as that for an ethylene plant, and some optimization packages, such as linear program packages, can be purchased from control system vendors or consulting firms. Even so, the time required to fit generalized models for a particular process can be costly in terms of time and effort. In addition, process constraints and economical models should be developed from actual operating data.

9.2.4 Economic Justification

Developing an economic justification for a supervisory control system, particularly for a complicated network system, can be difficult to quantify and may be very difficult to prove prior to the system's installation. Most of today's process plants can be operated with manual control, i.e., regulatory control with manually entered set-points. Distributed control systems at the lower level are justified because they are used to replace conventional controllers, sometimes at lower cost. Supervisory control and management information systems are not considered necessary to keep a plant running, but are necessary to maintain plant performance at a high level. Managers with conservative views frequently challenge the necessity of this control philosphy, and it is the job of the innovator to prove it is needed.

With few exceptions, the profit of a plant is just a small percentage of the total selling value of its products. In other words, profit is the difference between two large numbers, market value of products minus cost. Increased market value and reduced operating costs are the name of the game, and a small percentage increase in market value or decrease in operating costs will result in large percentage increases in profit; a one percent decrease in cost could result in a more than a 100 percent increase in net profit when cost and market value are nearly equal. The following paragraphs discuss potential justification for supervisory control systems in general, while some specific examples are given later.

9.2.4.1 Higher Plant Capacity

For a capital intensive plant, higher throughput capacity could be the best justification of all. A plant is designed with a "name-plate" capacity based on the averaged steady-state conditions, averaged quality of feedstocks, and averaged constraints. In reality, the operation of a plant rarely remains at the name-plate capacity. A plant may actually operate at an averaged capacity just as it was designed, but this hardly implies a real physical limitation.

Supervisory computers have been used to identify and expand the bottlenecks of a plant and to push the capacity to a new level. This is usually accomplished by a combination of better control and other debottlenecking projects such as adding a pump. The net results usually are significant increases in plant capacity, from a few percent to as high as 30 percent or greater (18). There have been many reports which try to give credit for capacity increases to each project, but one must recognize that without information from computer systems, many debottlenecking projects would never have been initiated. The realization of higher production capacity is the result of many engineering projects, but the computer project is usually the catalyst that initiates the activity.

Higher capacity should be considered as a major benefit even under so-called marketing-limited conditions. Because, even during the marketing-limited period, there could be a need for the higher capacity part of the time to respond to special orders. Flexibility in production capacity may determine the future business of a company.

9.2.4.2 Lower Operation Costs

A plant's variable costs consist of three major components. They are: (1) material and energy costs, (2) maintenance costs, and (3) operating labor costs. Supervisory computers can be used to reduce each of these costs. The following paragraphs explain how these savings may be achieved in specific plants.

Raw material and energy cost reduction Material savings are usually derived from improved yield. Improvements in yield are achieved by controlling a plant where a number of products are made from complicated chemical reactions. A typical example of this type of plants is an ethylene cracker where fresh feed, recycled hydrocarbons and steam are charged into a high temperature furnace to produce a gas containing a large number of coproducts. Depending on the process design, the fresh feed charged to an ethylene cracker can vary from light hydrocarbons, typically an ethane/propane cut, to heavier hydrocarbons approaching light gas oil. For a given process, the variation in feed quality, fuel heating value, and heat transfer through the furnace tubes due to variations in coke thickness are the major disturbances to the process, consequently, the operating conditions must be adjusted to compensate for these disturbances. Optimum operating conditions, such as hydrocarbon to steam ratio, feed rate, furnace temperature, depend on many factors. In a plant where conventional instruments are used, it is the operator's job to chase these moving targets by estimating the best setpoints for a number of controllers according to his understanding of the process. Computers have been installed to stabilize and optimize the operation of ethylene plants. The basic concept is to identify the disturbances by on-line measurements, and to determine the optimum operating condition according to process and optimization models. The computed results are used for closed-loop control. The operator's job is simplified, and improved product yields have been one of the major benefits (19).

Even where a process is not based upon a chemical reaction, the product yield may be improved via computer control. A typical example is a petroleum crude distillation unit. Figure 9.3 shows a

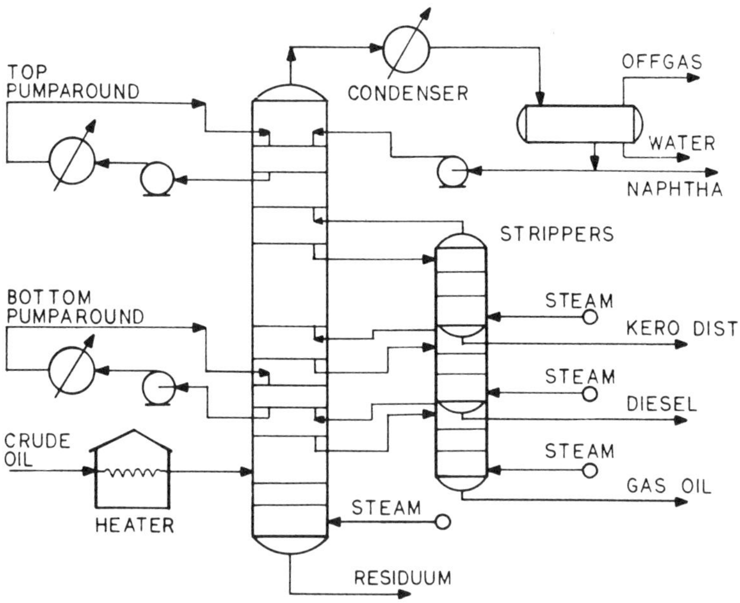

FIGURE 9.3 Crude distillation unit showing multiside draw product steam stripping, top and bottom pumparounds and charge heater for column temperature control and top product condenser and accumulator.

simplified process flow diagram for a crude distillation unit which is used to separate crude oil into several product streams with specified boiling ranges. Each stream, with the exception of the residuum, must meet quality specifications, and each stream also has a different market value. Quality specifications are set according to the intended use of the product. For example, should the kerosene be used as a jet fuel blending stock, it's "cloud point" becomes critical because jet fuel cannot tolerate the formation of solids at very low temperatures, such as −70°F. During manual operation, it would be the safe approach to keep the cloud point well below the specification. While this is safe from the product quality standpoint, it will also result in a lower yield of jet fuel stock, the difference being downgraded to light gas oil, which has less market value.

Computer control of the crude distillation unit will result in the production of jet fuel blending stock being maximized and the quality specification held within limits. This type of control is considered as an advanced regulatory function because the targets are already determined and the goal for control systems is to meet them (20).

Using the same example, maximizing one blending stock yield may not be the best economic goal under all marketing conditions. When the price difference of two products is small or the increased energy costs for better separation is large, it is possible that an optimum yield exists, and an on-line optimization function may be needed (17).

Energy savings can also be realized using computer control for major energy consuming equipment, such as boilers, furnaces, and kilns. Triggered by rapidly increasing energy costs in the late 1970s, there are now many successful computer control projects with the sole purpose of reducing energy consumption, and most of these have produced very positive results (21). In fact, there are control system vendors who offer turnkey projects to control boilers, heaters, etc., so the techniques are well-developed and the benefits widely recognized. However, energy savings do not necessarily come only from the control of well-known operations; just about any large energy consumer is a potential candidate for evaluation. One outstanding example is cement plant automation. The most important function for controlling a cement plant is probably the raw material blending. Good blending control is the key to stable kiln operation and consistent clinker quality. Kiln control itself is another important function, and the major goals for kiln control are to maintain desired throughput and product quality. Old time cement operators did not consider energy savings as a major goal, but energy savings can be indeed an important factor for a successful cement plant operation.

Figure 9.4 shows a simple process flow diagram of a cement kiln. The control of this kiln is a complicated task, but simply stated, it is achieved by manipulating the fuel flow to control the temperature profile of the solids passing through the kiln. Overfiring and underfiring are both undesirable conditions, but the process feedback is slow, so it is difficult to maintain stable operations. Computers have been used to control product quality, but energy savings have always been an important side benefit to provide economic justification for the project.

Even the performance of a very simple process can be improved by supervisory computer control. An example in this category is the optimization and control of cooling-water temperature in a cooling tower.

Figure 9.5 shows a simplified process control diagram for a cooling tower for process-cooling water. This process is so simple that many installations are not provided with measurement and control devices. All motor driven fans (M1, M2, and M3) are run continuously and the circulation pump is maintained at maximum load. The control of cooling-water temperature is usually ignored because it does not affect the operation of a plant so long as the temperature is below a certain critical value. But the control of cooling-water temperature can be important from a cost savings point of view. For a large plant, cooling fan motors can use hundreds of horsepower, and the electric power required to drive them can be costly.

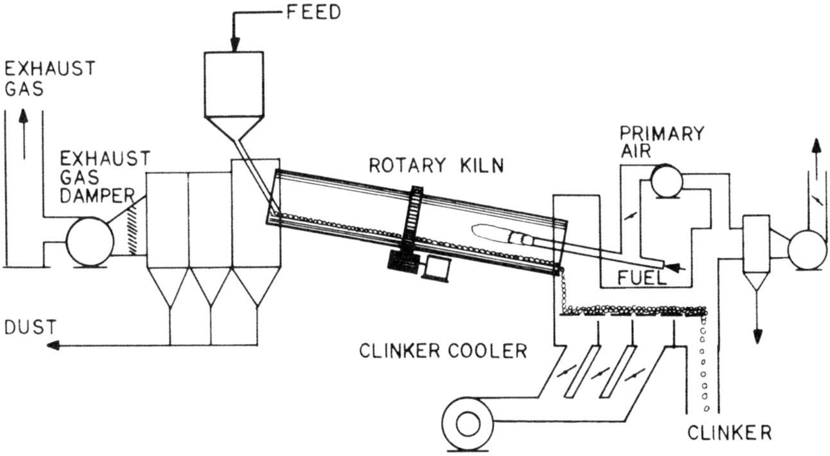

FIGURE 9.4 Rotary cement kiln for production of Portland cement with product-end grate cooler and heat recovery equipment and feed-end exhaust gas dust removal equipment and induced draft fan.

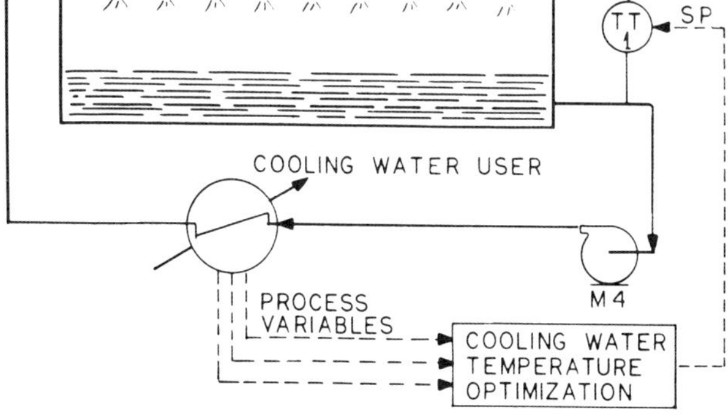

FIGURE 9.5 Cooling water tower with cooling-water temperature regulation and cooling-water use temperature optimization.

There can be two levels of control installed on a cooling tower. The function of the basic control level is to hold the cooling-water temperature at a setpoint by manipulating the cooling fan motors. The function of the supervisory control level is to determine the optimum cooling-water temperature which is defined as the temperature where the combined operating costs are at a minimum. The supervisory level function can be considered as a textbook example for a simple, practical optimization problem. The cooling fans are usually sized for the worst-case condition, that is, for hot and humid summer days. During that time, all cooling fans must be running, so there will be no control needed. When the ambient condition is not so hot and humid some strategy should be installed to set the cooling-water temperature.

Figure 9.6 shows the typical relationship between operating cost and cooling-water temperature. As the cooling-water temperature increases, the cost for operation of cooling tower fans decreases, but the cost for operating the equipment which uses the cooling water increases, and the combined cost has a minimum point. That point can be determined once the cost-temperature relationship is determined. Note that the minimum temperature must be somewhere above the wet bulb temperature of the air and is subject to the changes in operating conditions by all cooling-water users. Therefore, the minimum temperature achievable must be determined frequently.

In many cases, the optimization problem can be reduced to a minimization problem with a variable constraint condition. This is the

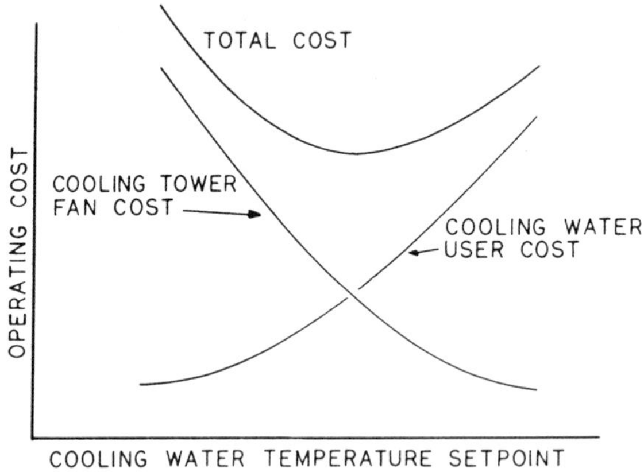

FIGURE 9.6 Operating curves showing cooling tower fan cost and cooling-water user cost as a function of increasing cooling-water temperature. The optimum operating point can be found by adding the two curves and locating the minimum point on the resulting total cost curve.

case when the increase in the operating cost of cooling-water users always exceeds the savings in that for the cooling fans. When this is the condition, the cooling-water temperature setpoint should be set according to the wet bulb ambient temperature. This will eliminate the waste of excessive cooling fan loads when the ambient temperature is low.

The above examples illustrate that energy savings can be realized from the control of both complicated and simple processes, and that for even a simple process, sophisticated control techniques may still be needed to generate the best possible savings.

Lower maintenance cost Maintenance cost savings are not generally considered as a direct benefit derived from supervisory control. Indeed, many managers are concerned about the cost of maintaining the control system, but do not consider the potential savings in maintenance costs as one justification for installing supervisory control computers. As a result, many important functions which may generate large savings in maintenance costs are not identified nor implemented.

As discussed in previous chapters, supervisory control systems can produce a more stable operation and also can analyze long-term trends in process variables. These are the two major functions which can be utilized to reduce maintenance costs. The following are some examples.

Figure 9.7 shows a multipass furnace designed for heating heavy hydrocarbons such as crude oil. Supervisory control systems can be used to balance all pass outlet temperatures and to monitor the furnace so that the temperature of any single pass will never exceed a predetermined value. Since the coking of the tubes is caused by high temperature, a balanced load and the elimination of high temperature upsets will reduce the chances for coking, thus, reducing the maintenance cost.

Another example is the monitoring and control of fixed bed catalytic reactors such as an isomerization unit in a petrochemical plant. This process unit is designed to convert paraffins into their isomers in a fixed bed reactor. This is a simple process as shown in Fig. 9.8. The catalyst in the reactor needs to be regenerated once it's activity is no longer sufficient to complete the reaction, and the regeneration of the catalyst is costly in terms of lost production.

Catalyst activity decay is influenced by two factors: poisoning and extremely high reaction severity. Both conditions are caused by process upsets. When the process is stabilized by supervisory control, the chances for both catalyst poisoning and extremes in reaction severity are reduced so maintenance costs can be reduced. Another important aspect for catalyst regeneration is to predict the time for regeneration, or better yet, to run the reactor until a scheduled plant maintenance time can be reached. The ability to use a computer to plot long-term trends of catalyst decay, and to

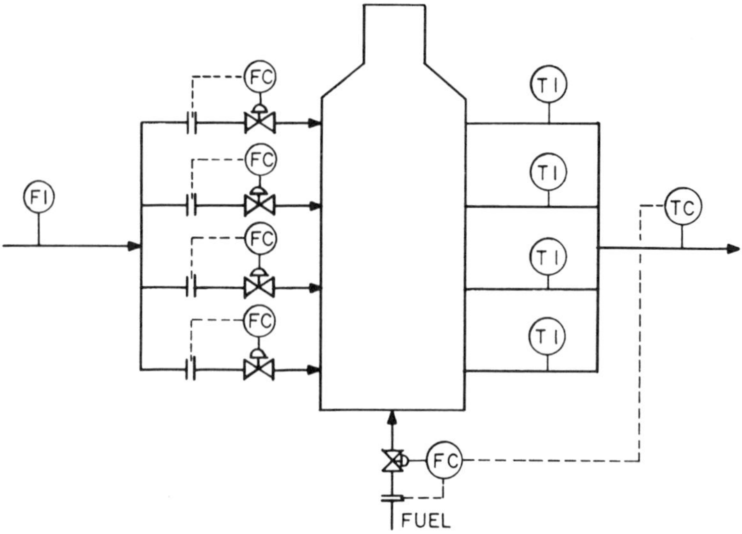

FIGURE 9.7 Multipass furnace with flow control and outlet temperature measurements for each parallel pass.

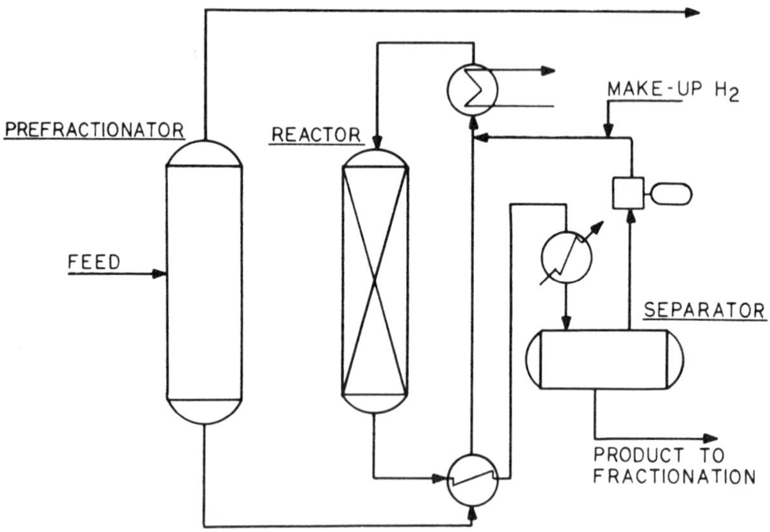

FIGURE 9.8 Isomerization unit for producing high octane gasoline blending stock showing prefractionator column, reactor, and product separator.

control the decay by allocating the loads to different reactors will be a great help in accomplishing these goals.

Lower labor costs savings and improve plant safety Most supervisory computers are not installed to save operating labor costs. In many cases, labor savings are not possible because of union rules or safety regulations. Under some conditions, labor savings are possible because of the consolidation of control rooms, and plant safety can be improved because automatic process monitoring functions are implemented.

Consolidation of control rooms has resulted from the application of advanced instrumentation technology. One centralized control room replacing several existing control rooms will require less direct operating labor even when there is little change in control strategy. This is accomplished by assigning more loops to be monitored by a single operator. There is no industrial standard as to how many loops an operator can supervise, but generally the number used is about 100 to 150 loops when conventional control strategy, i.e., single-variable controllers without supervisory computers, are used.

The actual number of loops an operator can supervise depends on the complexity of the process and the experience of the company, but the type of controller, whether analog or digital, does not make much of a difference. The reason is that when a digital controller is used just to replace an analog controller using the same control algorithm, the need for operator supervision is practically unchanged. When a supervisory computer is implemented, large numbers of single-variable loops are combined to create a few control strategies and many routine operator's tasks are automated, thus, one operator may supervise a much greater number of single-variable loops (22). Meanwhile, should automatic monitoring functions, as described in the last section, be implemented, operators would be warned about many abnormal conditions before these conditions became accidents.

Products having more consistent quality Although having products with more consistent quality, by itself, may or may not have a direct market value, it should be a goal for every manager, because products having more consistent quality always win market shares when everything else is equal. Supervisory control is probably the best existing technology to improve quality control, because the basic goal for supervisory control is to stabilize the process, and more stable processes always produce more consistent quality.

Better information An important by-product which results from the installation of a computer control system is information about the process being controlled. This information is valuable to the operator and the process engineer responsible for the operating performance of the process. In its most immediate form, the information is less meaningful to the plant management. However, when the information from the process control computer is combined with plant-wide information gained elsewhere, important management information can be developed.

The most commonly available information from control systems are cause–effect relationships between process variables. After a process is under computer control, engineers can be quite surprised by the insight provided into the cause–effect relationship which becomes clear when the process is stabilized. This information can be used to optimize the process, or it can even be used for improving future process design.

Another type of information by-product is a quantitative value for the performance of a process unit. A simple example is the thermal efficiency of a boiler or a heater, and a more complicated example is the conversion and yield of a catalytic reactor. This type of information is needed to identify operating costs in each unit and to plan how to improve the overall operating efficiency.

9.2.5 Feasibility Study and Management Support

Supervisory control projects have gained a strong reputation for being over budget as far as some corporate managers are concerned. Although very few written documents or statistics of actual installations support this reputation, word-of-mouth has indeed made many managers have doubts about the potential economic returns of a supervisory control project. It is not uncommon for a manager to approve a conventional plant improvement project with an estimated return on investment (ROI) at 20 percent per year, and then reject a supervisory control project with an estimated ROI at 30 percent per year. The uncertainty in cost estimation is usually given as the reason for such a decision. Based on the author's experiences, this has happened in several companies that have never implemented any supervisory control computers within their organizations, and yet the poor reputation is there. It was found that managers add large contingency factors to the estimated costs which already include reasonable continency factors, thus, greatly reducing the estimated ROIs.

Engineers must find some way to gain management confidence. The easiest way is to start a project whose cost can be firmly estimated. Examples of this are field proven turnkey systems, such as those available for paper machine control or for boiler optimization. Vendors can offer firm quotations for both system and installation costs. Since most technical problems should be solved by previous installations, the chances for project delay or cost overruns are extremely small. After some good track records like this, management may start accepting this type of project. This approach could be costly in the long run and will be discussed later.

For most processes, field-proven turnkey systems are not available. Some system vendors may take turnkey responsibility for any process, but it is not advisable to implement projects that way because desired performance cannot be well defined prior to the completion of the project and technical problems are not solved by previous projects. Furthermore, users may not have enough knowledge to enhance the system or even to maintain it, so the full potential of the system may not be reached. For this type of application, a supervisory control system needs to be developed instead of purchased as a total turnkey project, and a comprehensive feasibility study for the project must be made in order to gain management support in the long run.

A feasibility study should include all tasks described in the previous two sections, followed by an estimate of hardware and software, and implementation manpower requirements. It is important that this study be led by a person who has a strong background on the overall conceptual design and management of a supervisory control system. When such a person is not available within an organization, some outside help, such as a reputable consultant, should be used.

9.3 IMPLEMENTING A SUPERVISORY CONTROL PROJECT

Assuming the scope of a supervisory control project is determined and a project team is organized and staffed with qualified people, the next question to be addressed is how to start and manage a project of this type. This is a rather broad and often overwhelming question, and the answer is usually quite subjective, depending on one's experience (23,24). The authors believe the following generalized tasks should be considered at the beginning of a project.

9.3.1 Computer System Selection

A project engineer's approach to selecting a vendor is to generate a bidder's list, send out requests for bids, and award the contract to the lowest bidder meeting the specification requirements. For a computer system, this approach may work for some, but will not work well for many others. Generally speaking, it is extremely difficult to generate a sufficiently detailed specification which can really serve as a standard to compare apples with apples. When too much emphasis is placed on the proven system, one takes the risk of buying a more expensive and less capable system, and one may end up with obsolete equipment; when too much emphasis is placed on new technology, one takes the risk of debugging the system for the vendor which may delay project completion and increase the cost. The most prudent way to select a computer is to take a middle-of-the-road approach by separating development type projects from production type projects. The former are the projects where the main objective is to develop new application technology and, thus, "high-power" technical expertise is available, and the on-time completion of the project is not critical from the plant operation point of view. In this case, the most technically advanced system should be used. Note that this does not imply the use of expensive systems. On the contrary, many technically advanced systems carry low price tags and their application may require more technical know-how, but usually less capital investment than that for a more conventional system.

The production type of projects are those where on-time completion is important for the plant's operation, and the system is designed to implement proven functions. In this case, proven computers with proven standard software systems should be used. Recommended approaches for selecting computer systems for production type of projects follow:

9.3.1.1 Turnkey Projects

It is recommended that all turnkey projects have one important measurement that clearly defines the performance evaluation. A complete turnkey supervisory computer control system can be purchased for some special processes from specialized vendors.

An example of a complete turnkey system is the computer control of paper machines. Some vendors offer money-back performance guarantees for installing computers. Performance is measured by production increase and quality improvement. To select a system like this, one can use the low bidder approach, but should make sure that the system can communicate with other planned or installed supervisory computers without requiring extensive software work, and that the installation costs and the continuous service costs are included in the comparison.

The acceptance of complete turnkey systems, such as for paper-machine control within the pulp and paper industries, has been very popular because users do not have to provide the technical expertise to implement a system. In many cases they do not even need technical expertise to maintain the system. The vendor provides maintenance people on-site permanently to keep the system running. This is a simple approach to installing supervisory computers, but an expensive one in the long run because little or no application technology is developed within user companies. When the scope is broadened beyond the control of a special unit, these users may find that the lack of technical expertise within the company seriously restricts the implementation of future advanced control systems.

Some vendors also offer turnkey projects based on a more generalized performance criteria, such as where the hardware and standard software will meet a certain performance standard. Such a system can be considered as part of an in-house project which will be described next.

9.3.1.2 In-House Projects

An in-house project means that the computer system is specified and purchased by the user and is implemented by the user, but it does not necessarily imply that subcontractors are not used. In fact, subcontractors should be used whenever packaged systems can be applied, but the responsibilities of subcontractors should be well defined and their efforts properly coordinated during implementation by the user's project team. For in-house projects, users have great flexibility in the selection of both the system hardware and software, but the decision must be made with careful technical and economical consideration.

Hardware configuration The classical approach for developing a hardware configuration for a digital control system was to specify one or two large computers to implement all of the desired functions. The reasoning was that computers were expensive and communication between computers was difficult, somewhat uncertain, and expensive. Technological advancement in recent years have changed all of this. One can still purchase a very large computer to implement all desired functions, but one can also use a number of smaller computers combined with a high-speed data communications network to link them together, such as the configuration shown in Fig. 9.1.

For a user without previous experience in on-line computer applications, the following considerations should be made for selecting hardware systems:

Identify those functions which can be implemented by packaged hardware and software systems, such as computer-driven chromatographs, environmental monitoring systems, operator training computers, etc.

Select computer systems which have been used for similar applications by others with positive results, but not a system which is already technically obsolete.

Select a system which offers advanced self-diagnostic capability, such as facilitating troubleshooting at a remote location via telephone communications.

When more than one computer is used, it is preferable to use the same model, or different models from the same vendor. When this is not feasible because of other considerations, make sure each system selected is compatible with the overall communications network.

All computers, with the exception of those purchased as part of the complete turnkey systems, should support a common high-level language such as FORTRAN.

Try to use dedicated computer for software development, although this task can be accomplished by using the "background" programming capability of a control computer, it is much simpler from the software point of view to use a dedicated computer. Note this computer should be connected to the communication network also.

Software configuration As discussed in Chapter 5, computer software can be categorized as either standard software or application software. Standard software includes those programs that are purchased with the computer hardware. These programs provide a structure for the user to support his application software development. Alone, the standard software packages have little functionality from the user's standpoint until the applications software programs are added. Application software is a series of programs which tells the computer what particular functions the user wants his computer to perform. For example, a FORTRAN Compiler is a standard program. A program written in FORTRAN language to generate a CRT display is an application program.

Many computer vendors or software system houses offer generalized application software packages, which are programs designed for multiple applications. Users may take a generalized application package and implement their desired functions by selecting a few parameters and entering a few simple codes. This can be demonstrated using an example for the generation of a CRT display. Typical standard software from a computer vendor for this purpose consists of only the CRT driver, which gives a standard subroutine to display information on a CRT screen. A programmer must then write a program, using whatever language plus those subroutines, to generate the display. A programmer may spend several days to generate one display, and several more days to generate another display. A generalized CRT package offers programs that will greatly reduce the program development effort. This package allows users to use a predefined procedure to generate the CRT display without writing a program. Normally, an engineer can learn how to use this system in a few days, and once familiar with the system, can generate displays in a few hours.

Manpower savings derived from using generalized application software packages is only one advantage. An even more important advantage is that engineers can implement desired functions without requiring a programmer to interpret their ideas. Ease of software maintenance is also accomplished because of the reduction in the need for program coding.

Generalized application packages have been developed for many purposes other than CRT display generation. The most commonly used types are generalized process control software packages (25). To use this type of package, a user is only required to fill-in-the-blank to generate his system. Thus, this type of system is also called a "fill-in-the-blank" software system, and can usually be used to provide a good starting point for implementing computer control projects with greatly reduced overall software development costs.

It is not practical to discuss the detailed pros or cons of the application of fill-in-the-blank packages because of the in subjective nature. Based on the authors' experiences, this type of software package can be applied to accomplish at least 70 percent of desired tasks, with the balance accomplished by special programs using high-level languages. This makes the use of fill-in-the-blank systems economically feasible.

The following factors should be considered when a fill-in-the-blank system is selected:

Software packages may include some control algorithms, but the task of control analysis is the responsibility of the user-engineer. Do not use an algorithm without fully understanding it and rationalizing the use of it.

Try to define the control functions, including operator communications, so that they are not dependent on bulk memory devices such as discs. When bulk memory devices must be used, specify the random access memory (RAM) to be as large as possible to minimize the frequency of bulk memory accesses.

Use a floating point data format for all programs. Some systems were designed using fixed-point data formats to save memory and to reduce computation time, but fixed-point data manipulation creates many difficulties, and should be considered unacceptable.

All data acquisition and control functions are basically organized as loops and these loops are processed according to predefined time cycles. It is important to keep the period of each loop precisely controlled, independent of the computer computational load.

All control loops should have the ability to turn on user written programs.

The software system should allow a user to test a program using on-line data, but should not allow that program to change any location outside its own memory area.

The software package support program documentation.

There are many more factors to be considered, such as the ease of system generation, the availability of standard operator communications programs, etc. There are also nontechnical factors such as the overall reputation of the vendor and his potential for staying in business for the next ten years. A user company should evaluate his resources and needs before selecting a software system. Once a system is selected, it could be difficult and expensive to change to another one.

9.3.2 Implementation Management

Contrary to the beliefs of some people, supervisory control projects can be planned and scheduled just like any other engineering project. The only special characteristic of a supervisory control project is that milestones are hard to define and a project team is likely to use all of the available time and money allocated, mainly because a software task can always be improved. In order to control a project properly, project milestones must be defined during functional specification time by considering manpower requirements for all hardware and software development tasks. Just meeting simple milestones such as on-site hardware delivery will not assure the project is progressing well.

To develop a project schedule and assign priorities for all tasks, one should consider the following characteristics of a computer system. The most important, or beneficial, task may not require the largest share of resources. The manpower and computer memory requirement to implement a control function is usually much less than that needed to implement a special CRT display to support that control function. Many supporting functions are in the nice-to-have category and should be assigned a lower priority than that for the must-have functions.

People always gain new information from the activity of implementing a control system. This new information may prove that some of the functions defined in the specification are not feasible. A periodic review of the purposes of the specified functions should be made, and the unnecessary or impractical functions should be eliminated or replaced by others.

Perhaps the most special characteristic about managing a supervisory control project is the need for keeping consistent project team members during the project's entire duration. This is mainly due to the long learning period required to become an effective contributor to the project, particularly for the people who need to use the software system. When an engineer or a programmer is assigned to a project for six months and then replaced by another person, it is most likely that the six man-months effort by the first person is wasted and must be repeated. The computer system has been changing so fast that few people are likely to use systems with identical design for two consecutive projects, and the time required to become proficient on a computer system is about six months for an experienced person.

New supervisory control systems will be developed with the advancement of new technology. It is not practical to predict what form of computers will be used in the future, but one thing is certain, there will be change. The authors believe that the best way for engineers to cope with these changes is to keep up with new technologies and also apply them with general engineering common sense. This, after all, is what engineering is all about.

APPENDIX A: IMPLEMENTING BASIC AND ADVANCED CONTROL FUNCTIONS USING A HONEYWELL SYSTEM

A.1 The Purpose

This section provides an illustration of an actual implementation of a multilevel control strategy using the Honeywell TDC-2000 system and a Honeywell 4500 computer. The following basic implementation concepts will be demonstrated by this illustration.

1. The separation of basic control functions and advanced control (regulatory) functions.
2. The importance of a user friendly advanced control function, i.e., automatic initialization, etc.
3. The use of the actual control system as a process simulator.

The methods for setting up a Honeywell TDC-2000 basic control loop and how to use a BICEPS loop will also be discussed.

A.2 The Process and the Control System Design Concept

The multipass furnace is a typical process operation found in a petroleum refinery and was selected for this illustration. This is shown in Fig. A.1. The desired control functions for this process are:

1. Heat the combined feed to a desired temperature (set by operator).
2. Maintain the temperatures of all passes as close to each other as possible.
3. Increase or decrease the feed rate without disturbing the temperatures.
4. Change the fuel allocation without disturbing the temperatures.

A conventional control system, using analog controllers, only controls the first function listed above automatically; all other functions are

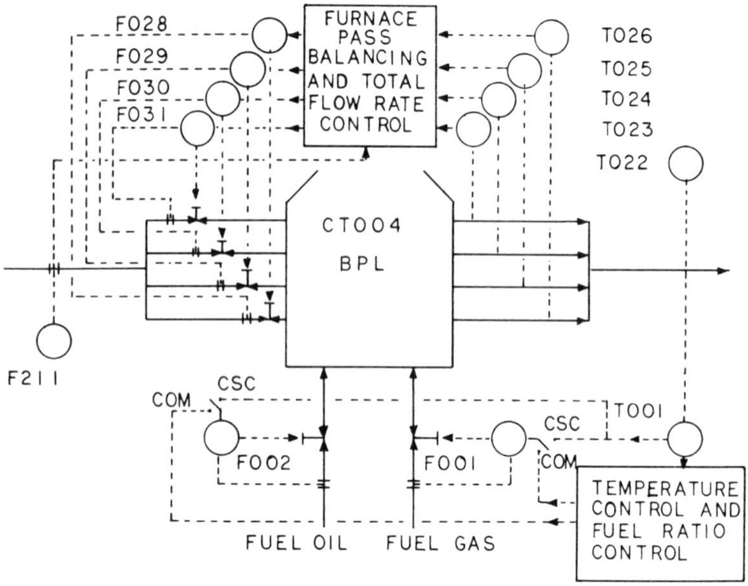

FIGURE A.1 Four-pass furnace with two full sources.

accomplished by manual control. Operators are familiar with this type of control system and can maintain the process at stable, but not optimum conditions.

In order to minimize the temperature differences between passes, and to minimize disturbances to feed temperatures, an advanced control system can be designed to automate functions (2) to (4) listed above based on the following criteria:

Operators can still run the process as they used to.
Operators can turn on or off the advanced control functions with ease.
Advanced control systems can be designed to identify all potential equipment failures and be fail-safe.

Honeywell's TDC-2000 basic controllers were used to replace all existing analog controllers and a Honeywell 4500 computer with PMX software was used to implement all advanced control functions. As shown in Fig. A.1, all basic control functions are indicated by circles, while all advanced control functions are indicated by rectangles.

In order to train the operators on the operation of the new control system for such tasks as calling up CRT displays and closing advanced control loops, a process simulator was designed and implemented using the Honeywell 4500 computer. Figure A.2 shows the hardware configuration utilized for the simulator, and Figure A.3 shows the information flow diagram for the control system and the simulator. Note that the simulated process outputs are sent to a basic controller which is used as a sample and hold device.

The simulator initialization function block shown in Figure A.3 was designed to reduce the efforts to set-up the simulator. Engineers can bring the process and the control system to a predetermined operating condition by executing this program. The concept for using a computer program to perform condition checks and other initialization actions has proven to be very successful for developing a user friendly design system.

Process models are used to demonstrate process dynamics and interactions when the process is disturbed, thus, no precise steady-state data are needed. Models can be developed with limited effort. For example, a flow process is simulated by a first-order filter as shown in Figure A.4; the temperature of a heater pass decreases with increasing load, when everything else is held constant.

A.3 Implementation Procedures

Implementation has been achieved utilizing the basic controllers and computer as follows:

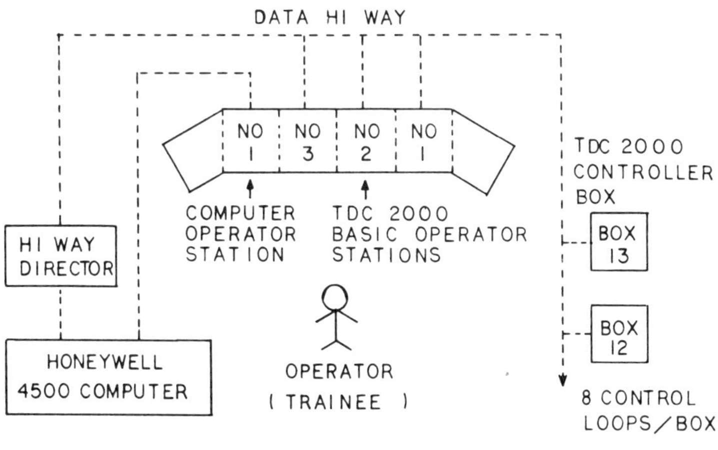

FIGURE A.2 Hardware configuration for the operator training simulator.

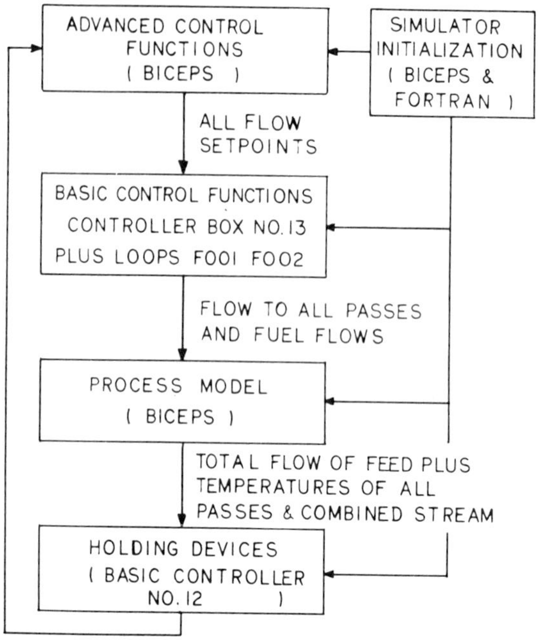

FIGURE A.3 Information flow diagram.

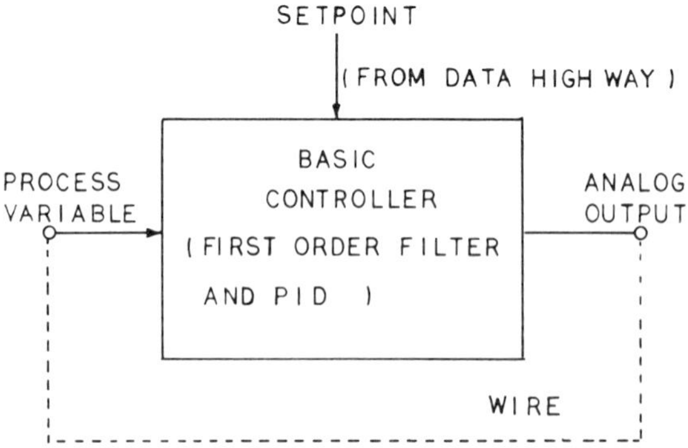

FIGURE A.4 Wiring of basic controller for testing and simulation (without independent simulator).

1. TDC-2000 Basic Controller

The following controllers can be manipulated by operators when they are switched to *manual* or *auto* mode, and they can accept computer commands when they are switched to computer *comp* mode.

1. F031 pass 1 flow controller
2. F030 pass 2 flow controller
3. F029 pass 3 flow controller
4. F028 pass 4 flow controller
5. F001 fuel oil flow controller
6. F002 fuel gas flow controller
7. T001 furnace outlet temperature controller

The following controllers are used to simulate the sensing devices attached to the process. Analog signals from the furnace model are "held" in these loops to simulate the actual furnace response. These loops are:

Loop ID	Description
T026	Pass 1 outlet temperature
T025	Pass 2 outlet temperature
T024	Pass 3 outlet temperature
T023	Pass 4 outlet temperature
T022	Combined outlet temperature
F211	Measured total feed flow

The detailed controller configurations are listed in Table A.1.

2. Computer System

The process simulation is accomplished using the mathematical model which is derived from mass flow and energy balance equations. These equations were programmed using the BPL language, which is similar to the familiar FORTRAN language, and the resulting BPL program was assigned to the loop ID CT004 for periodic execution. This model computes and generates responses to changes made by operators, by the supervisory control, or by process disturbances (such as fuel heating value changes).

For advanced supervisory control, the control strategies are subdivided into four major loops for computer control.

TABLE A.1 Controller Configuration Detail

Box No.	Slot No.	Conf	Conf Hi	Conf Lo	Conf Hi Lo	ID	Eng. Unit	Description
13	1	0701	1113	0001	0010	F028	BPH	Pass 1
13	2	0701	2123	0001	0010	F029	BPH	Pass 2
13	3	0701	3133	0001	0010	F030	BPH	Pass 3
13	4	0701	4143	0001	0010	F031	BPH	Pass 4
13	7	0702	7173	0001	0010	F211	BPH	TOT Flow
12	1	0701	1113	0001	0010	T026	DEG F	Pass 1
12	2	0701	2123	0001	0010	T025	DEG F	Pass 2
12	3	0701	3113	0001	0010	T024	DEG F	Pass 3
12	4	0701	4143	0001	0010	T023	DEG F	Pass 4
12	5	0701	5053	0001	0010	T022	DEG F	Cmbn Temp
12	6	0501	5063	0001	0010	T001	DEG F	Out Temp
12	7	0713	7160	2001	0010	F001	BPH	Fuel Flw
12	8	0701	8160	2001	0010	F002	MSCF	Gas Flw

Loop ID	Function
CT002	Perform the pass temperature balancing control
C211	Perform total flow control with ramping capability
CT003	Perform feedback and feed-forward control
CR001	Perform the fuel ratio control for minimum fuel cost control

These loops are implemented in the computer as shown in Fig. A.5. The detailed interconnections of total flow control (C211) and temperature balance control (CT002) are shown in Fig. A.6. There are five process input variables: the total flow (F211) and four pass outlet temperatures—T026, T025, T024, and T023. There is one operator input for total flow setpoint (C211 sp). The outputs of these loops command individual pass flow controller setpoints—F031 sp, F030 sp, F029 sp, and F028 sp.

The detailed implementation of outlet temperature control and fuel ratio control is shown in Fig. A.7. There are two process input

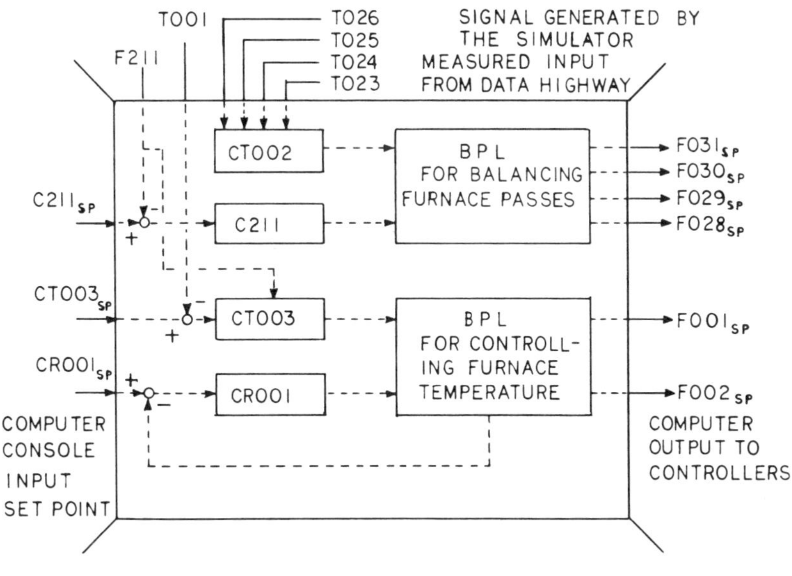

FIGURE A.5 BICEPS loop organization for furnace control strategy (implemented in computer).

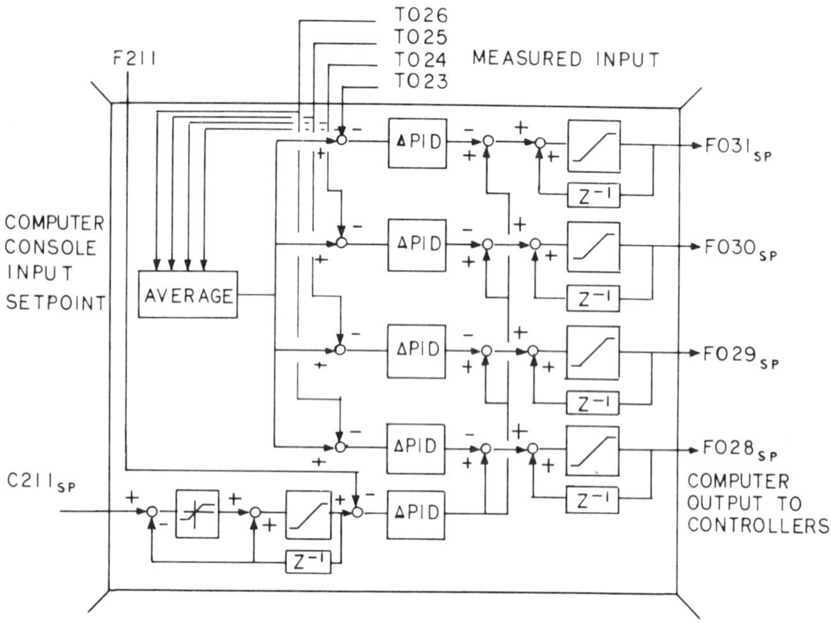

FIGURE A.6 Total flow control with ramping and temperature balance control.

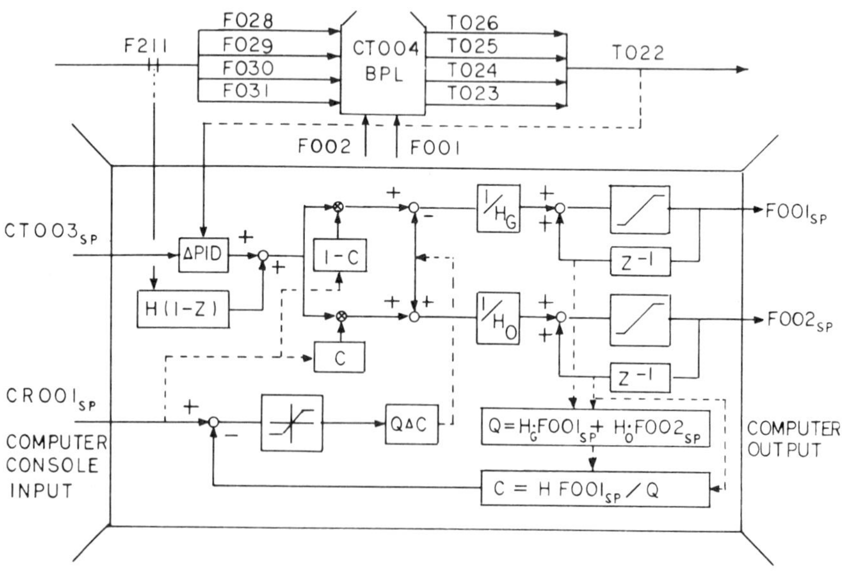

FIGURE A.7 Outlet temperature control and fuel ratio control.

variables. One is the total flow rate (F211) and the other is the combined outlet temperature T022 (both are computed by the process model and output to TDC-2000 during simulation or measured by on-line sensors during actual operation). There are two operator console input setpoints, CT003 sp and CR001 sp; and two computer outputs for fuel oil flow and fuel gas flow setpoint commands, F001 sp and F002 sp.

APPENDIX B: DISTILLATION COLUMN CONTROL SIMULATION

B.1 How to Simulate a More Complicated Process?

The hardware system configured in Fig. A.2 of Appendix A can be used to simulate any desired process and to train operators. This section shows the method for simulating a process which is more complicated than the four-pass furnace discussed in the previous appendix, that is, a distillation column. The basic design concept for the simulator is identical to that for the furnace simulation, i.e., using the TDC-2000 controllers to simulate basic control and to hold computer outputs, and to use the computer to simulate advance control and process responses. However, the implementation procedure is somewhat different. The model for the distillation column is programmed using FORTRAN language instead of BPL as in the last example. The reason for this change is that BPL programs can use only a limited amount of computer memory, which is too small for simulating a distillation column. In addition, not all of the process model outputs are held in basic controllers, as for the last simulation model, because variables such as product analysis are not available at the basic operator console in the actual installation, nor are they available in the simulator's basic console. (The values of the furnace outlet temperatures for the last example must be held in basic controllers so they can be displayed at the basic console. For actual installation, these temperatures are measured variables which are available to the basic control.) All of the computer model outputs are displayed at the computer console and are available for advanced control.

B.2 Distillation Column Model

A distillation column with single feed stream and hot oil reboiler is simulated. A process and instrumentation diagram is shown in Fig. B.1. The feed consists of five components; they are Propane, *i*-Butane, *n*-Butane, *i*-Pentane, and *n*-Pentane. The input variables and the output variables to and from the process model are shown in Fig. B.2.

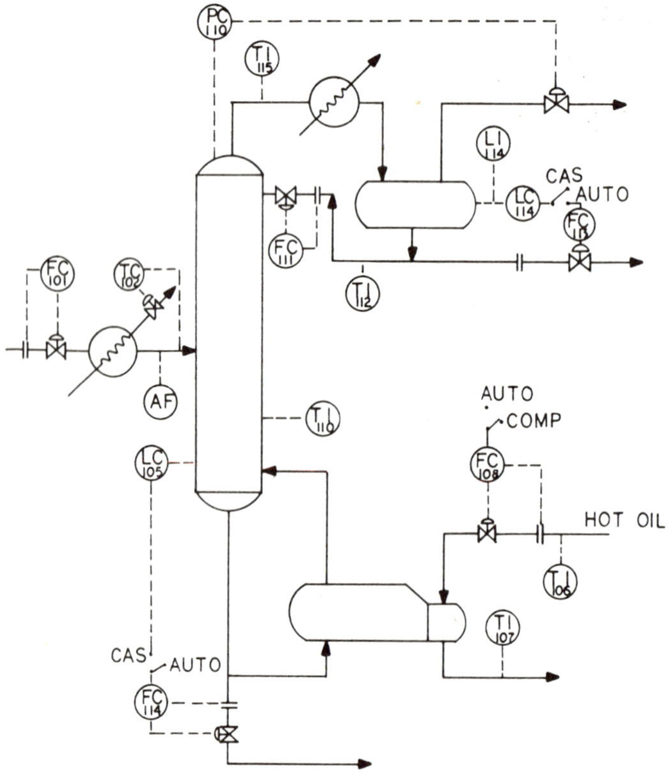

FIGURE B.1 Typical control instrumentation and process flow diagram for distillation column control.

The basic controllers, which can be manipulated by the operator through setpoint change, are:

ID Name	Description
FC101	Feed flow
TC102	Feed temperature
FC108	Hot oil flow
FC111	Reflux flow
FC113	Distillate flow
FC114	Bottom flow

ID Name	Description
FC110	Column pressure
LC114	Overhead accumulator level
LC105	Column bottom level

The disturbance inputs, which can be adjusted by the instructor for training the operators, are:

ID Name	Description
INPC1	Percent Propane in feed
INPC2	Percent *i*-Butane in feed
INPC3	Percent *n*-Butane in feed
INPC4	Percent *i*-Pentane in feed
INPC5	Percent *n*-Pentane in feed
TI106	Hot oil temperature to reboiler
TI107	Hot oil temperature leaving reboiler

All of the variables listed above were entered into the model periodically, and the following process responses are computed by the model:

ID Name	Description
TI115	Overhead temperature
TI110	Bottom temperature
TI112	Reflux temperature
LC114	Accumulator level
ADC1	Distillate Propane percent
ADC2	Distillate *i*-Butane percent
ADC3	Distillate *n*-Butane percent
ADC4	Distillate *i*-Pentane percent
ADC5	Distillate *i*-Pentane percent
LC105	Bottom level
ABC1	Bottom Propane percent

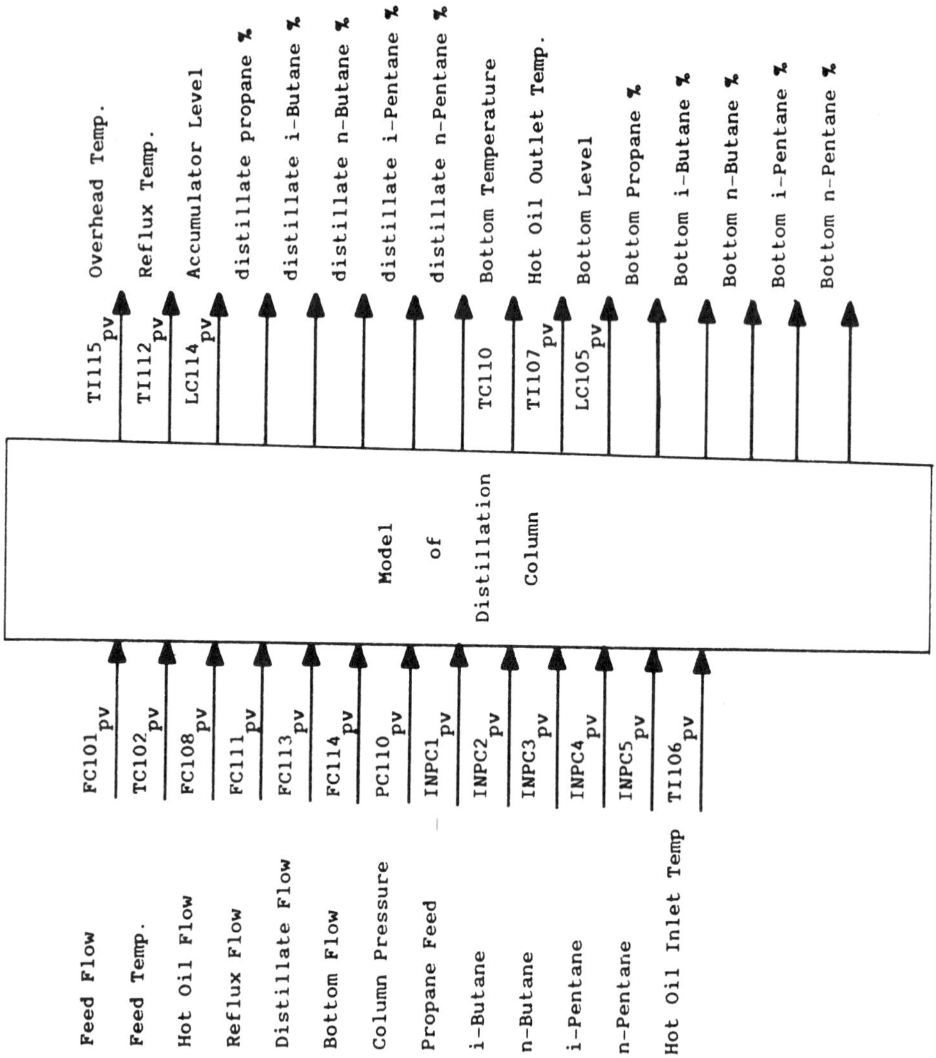
Feed Flow
FC101pv
Feed Temp.
TC102pv
Hot Oil Flow
FC108pv
Reflux Flow
FC111pv
Distillate Flow
FC113pv
Bottom Flow
FC114pv
Column Pressure
PC110pv
Propane Feed
INPC1pv
i-Butane
INPC2pv
n-Butane
INPC3pv
i-Pentane
INPC4pv
n-Pentane
INPC5pv
Hot Oil Inlet Temp
TI106pv
Model of Distillation Column
TI115pv
Overhead Temp.
TI112pv
Reflux Temp.
LC114pv
Accumulator Level
distillate propane %
distillate i-Butane %
distillate n-Butane %
distillate i-Pentane %
distillate n-Pentane %
TC110
Bottom Temperature
TI107pv
Hot Oil Outlet Temp.
LC105pv
Bottom Level
Bottom Propane %
Bottom i-Butane %
Bottom n-Butane %
Bottom i-Pentane %
Bottom n-Pentane %

ID Name	Description
ABC2	Bottom *i*-Butane percent
ABC3	Bottom *n*-Butane percent
ABC4	Bottom *i*-Pentane percent
ABC5	Bottom *n*-Pentane percent

The distillation model assumes optimum feed tray location, a total condenser, and a partial reboiler. It uses the Fenske and Underwood shortcut method as described in *Design Equilibrium Stage Processes* by Buford D. Smith (26). The thermodynamic data were taken from Perry's *Chemical Engineering Handbook* (27). Process dynamics were simulated by adding time delays and first order filters to calculated results.

As discussed in Appendix A, all simple feedback and cascade control functions were implemented using the Honeywell TDC-2000 basic controllers, and all advanced functions (mostly multivariable controllers) were implemented by using the Honeywell 4500 computer. Operators can practice control of the process in either manual mode, auto mode, cascade mode, or computer mode.

It was found that the basic design concepts of separating advanced control from basic control, and of implementing user friendly operating procedures are well accepted by operators, and the use of a simulator like the two examples discussed above for operator training is indeed worth the effort.

B.3 Discussion

One word of caution must be made here. The simulators described here are designed for the operator's training during normal operation, but not for process start-up and shutdown. There are commercially available simulators based on mini- or microcomputers designed for that purpose. When a complete training package is needed, it could be more economical to purchase a commercially available package. However, when the purpose is to give operators a basic understanding of the control system, or more importantly, to test the in-house developed advanced control functions, simulators such as those described here should be developed. One side benefit from the development of these systems is to give engineers a chance to practice the

FIGURE B.2 Input/output of distillation model.

implementation procedure and have some feedback from the system which is to be used to control the plant. This is usually a very valuable experience.

REFERENCES

1. Roberts, S. M. and Laspe, C. G., On-line computer control of thermal cracking, *Industrial and Engineering Chemistry, 53*(5):343–348 (May, 1961).
2. Shah, M. J. and Stillman, R. E., Computer control and optimization of a large methanol plant, *Industrial and Engineering Chemistry, 62*(12):59–75 (December, 1970).
3. Webb, P. U., Steward, W. S., and Smith, D. E., Process computer applications of refinery units, 1976 NPRA Computer Conference, Philadelphia, Pennsylvania, November, 1976, pp. 1–8.
4. Holt, A. B., Improved fractionation thru computer control, *Instrumentation Technology, 18*(1):43–48 (January, 1971).
5. Laduzinsky, A. J., Software takes modular forms for effective process and factory control, *Control Engineering, 29*(11):64–67 (November, 1982).
6. Haggin, J., Process control on way to becoming process management, *Chemical & Engineering News, 62*(19):7–13 (May, 1984).
7. Baqui, Abdul, The Computer Mystique: Let's put it to rest, *Instrumentation Technology, 32*(6):31 (June, 1985).
8. Nisenfield, A. E., Better process performance: Not on the data sheet, *Instrumentation Technology, 30*(2):19–20 (February, 1983).
9. Nasi, M., Souvander, M., Tuomala, M., White, D. C., and Darby, M. L., Experience with ethylene plant computer control, *Hydrocarbon Processing, 79*(6):74–82 (June, 1983).
10. Williams, T. J., Hierarchical distributed control: Progress and needs, *Instrumentation Technology, 30*(3):45–52 (March, 1983).
11. Moore, R. L., Adding real-time expert system capabilities to large distributed control systems, *Control Engineering, 32*(4):118–121 (April, 1985).
12. Chowburg, J., Process controllers don "Expert" guises, *Chemical Engineering, 92*(13):14–17 (June, 1985).
13. Pugh, L. C. and Cook, G. J., Human factors in the design of an advanced control complex, Proceedings of the ISA International

Conference and Exhibit, Houston, Texas, October, 1984, pp. 1397–1410.

14. Funk, G. L. and Kania, C. C., Optimizing and entire olifins plant pays off, *Oil and Gas Journal*, *82*(36):74–78 (September, 1984).

15. Beveridge, G. S. G. and Schechter, R. S., *Optimization: Theory and Practice*, McGraw-Hill, New York, 1970.

16. Douglas, J. M., *Process Dynamics and Control*. Englewood Cliffs, New Jersey, Prentice-Hall, 1972.

17. Kane, L. A., *Process Control and Optimization Handbook*. Gulf Publishing, Houston, Texas, 1980.

18. Funk, Gary L. and Smith, Dexter E., Optimizing fractionation performance through advanced control, *Instrumentation Technology*, *28*(8):35–39 (August, 1981).

19. Asgari, M., Yang, C. H., Minarik, Z., and Zelenka, J., Computerization of a modern ethylene plant, *Chemical Engineering Process*, *79*(11):27–33 (November, 1983).

20. Rinne, R., Sunnel, H., Latour, P. R., and Payntex, K. K., Experience with distillation unit computer control, *Hydrocarbon Processing*, *61*(3):141–148 (March, 1982).

21. Soderstrom, E. D., Computer control for energy savings, *Chemical Engineering Progress*, *76*(8):60–62 (August, 1980).

22. Quarin, S. A., An Attempt to Harmonize Operator and Computer Responsibilities in Complex Industrial Processes, Honeywell PMS User Group Symposium, Phoenix, Arizona, June, 1982.

23. Inoue, Y., Practical Management and Control Systems in the Steel Industry—A Case Study of Kimitsu Integrated Information System, Case Study-5, IFAC Congress, August, 1981.

24. Ben-Yaacov, Gideon, Effective Project Management—The Key for Successful Implementation of Advanced Digital Control and Monitoring Systems, Proceedings of the ISA International Conference and Exhibit, Houston, Texas, October, 1984, pp. 355–365.

25. Alexander, J., Software Packages for Process Control, Proceedings of the ISA International Conference and Exhibit, Houston, Texas, 1984, pp. 641–648.

26. Smith, Buford, D., *Design of Equilibrium Stage Processes*, McGraw-Hill, New York, 1963.

27. Perry, Robert H., and Green, Don, *Perry's Chemical Engineering Handbook*, McGraw-Hill, New York, 1984.

Index